SpringerBriefs in Earth Sciences

For further volumes:
http://www.springer.com/series/8897

Mike Fuller

Our Beautiful Moon and its Mysterious Magnetism

A Long Voyage of Discovery

 Springer

Mike Fuller
Institute of Geophysics and Planetology
University of Hawaii at Manoa
Honolulu, HI
USA

ISSN 2191-5369 ISSN 2191-5377 (electronic)
ISBN 978-3-319-00277-4 ISBN 978-3-319-00278-1 (eBook)
DOI 10.1007/978-3-319-00278-1
Springer Cham Heidelberg New York Dordrecht London

Library of Congress Control Number: 2013941497

© The Author(s) 2014
This work is subject to copyright. All rights are reserved by the Publisher, whether the whole or part of the material is concerned, specifically the rights of translation, reprinting, reuse of illustrations, recitation, broadcasting, reproduction on microfilms or in any other physical way, and transmission or information storage and retrieval, electronic adaptation, computer software, or by similar or dissimilar methodology now known or hereafter developed. Exempted from this legal reservation are brief excerpts in connection with reviews or scholarly analysis or material supplied specifically for the purpose of being entered and executed on a computer system, for exclusive use by the purchaser of the work. Duplication of this publication or parts thereof is permitted only under the provisions of the Copyright Law of the Publisher's location, in its current version, and permission for use must always be obtained from Springer. Permissions for use may be obtained through RightsLink at the Copyright Clearance Center. Violations are liable to prosecution under the respective Copyright Law.
The use of general descriptive names, registered names, trademarks, service marks, etc. in this publication does not imply, even in the absence of a specific statement, that such names are exempt from the relevant protective laws and regulations and therefore free for general use.
While the advice and information in this book are believed to be true and accurate at the date of publication, neither the authors nor the editors nor the publisher can accept any legal responsibility for any errors or omissions that may be made. The publisher makes no warranty, express or implied, with respect to the material contained herein.

Printed on acid-free paper

Springer is part of Springer Science+Business Media (www.springer.com)

*To my dear Patricia, whose love, courage
and strength have been an inspiration for me*

Acknowledgments

I thank those in my family, who made my life in science possible, particularly my uncle Johnnie Clegg and my Aunt Marjorie for encouraging me on that path. Thanks also to those at Christ's Hospital and Cambridge and helped me on my way and to my many friends in the academic world with whom I have shared my journey.

Thanks to Natalia Bezaeva, Jim Dewey, Susan Halgedahl, Chris Harrison, Bill Lowrie, Bob McClone, Ed Scott, and Ben Weiss, who were kind enough to look at parts of early versions of the manuscript and made many helpful suggestions.

Many thanks to Ben Weiss and his research group at MIT for including me in their efforts and sharing so many important ideas. In particular, I would like to thank to Sonia Tikoo from MIT, and Jérôme Gattacceca and Pierre Rochette from the group at Aix en Provence, for their help. I am also most grateful to my friends at the University of Hawaii at Manoa, especially Ed Scott, Pavel Zinin amd Jeff Taylor with whom I have had such good discussions.

I thank Annette Buettner, Janette Sterritte and Dr. S.A. Shine David at Springer Verlag for their interest in this manuscript and help. Annette Buettner has been particularly kind and helpful in shepherding me through the publishing process.

Finally, I thank my daughter Karen for her help with the diagrams.

Contents

1	**The Moon in Antiquity and in the Development of Modern Science**	1
	1.1 The Moon in the Greek World	1
	1.2 The Beginning of the Age of Modern Science	9
	1.3 Evolution of the Earth Moon System: Tides and Precession	12
	1.4 Summary	14
2	**Lunar and Earth Sciences at the Time of the Apollo Landings**	15
	2.1 Early Modern Studies of the Moon and Impact Cratering	15
	2.2 Geology and Geophysics in the 1960s	17
	2.3 Summary	21
	References	21
3	**The Birth of the Space Age and Unmanned Missions to the Moon**	23
	3.1 First Steps into Space and Early Heroes	23
	3.2 Unmanned Missions to the Moon and the Beginnings of Apollo Program	29
	3.3 Summary	31
	References	32
4	**Apollo: Getting to the Moon**	33
	4.1 Apollo Background	33
	4.2 Apollo 1 to 11	37
	4.3 Brief Summary of Remaining Apollo Missions	41
	4.4 The Training of Apollo Astronauts as Geoscientists	43
	4.5 Summary	45
5	**Advances in Lunar Science with Apollo**	47
	5.1 Apollo Scientific Results	47
	5.2 Lunar Stratigraphy	51

	5.3	Paleomagnetism of Apollo Samples	52
	5.4	Magnetism of Lunar Crust: Surface and Sub-satellite Observations	54
	5.5	Summary	56

6 The Earth's Magnetism: Paleomagnetism as a Rosetta Stone for Earth History ... 57

	6.1	The Geomagnetic Field and Its Paleomagnetic Record	57
	6.2	Magnetic Fields of the Cosmos	62
	6.3	The Paleomagnetic Record as a Rosetta Stone for Earth History	63
	6.4	Paleomagnetism: How the Magnetic Field is Recorded by Rocks	63
	6.5	Summary	66
		References	66

7 Lunar Paleomagnetism: Methods and Preliminaries ... 67

	7.1	The Continuing Debate Over Lunar Magnetism	67
	7.2	Laboratory Techniques	69
	7.3	Lunar Paleomagnetism: Beginning to Attack the Puzzle	70
	7.4	Summary	72
		References	72

8 Impact Related Shock on the Lunar Surface and the Lunar Paleomagnetic Record ... 73

	8.1	Cratering on the Earth and Moon	73
	8.2	Shock Effects on Remanent Magnetization	76
	8.3	Magnetic Fields of Terrestrial Impact Craters	82
	8.4	Summary	82
		References	83

9 Lunar Paleomagnetism and the Case for an Early Lunar Dynamo ... 85

	9.1	Criteria for Reliable Recording of Ancient Lunar Fields by NRM	85
	9.2	Apollo 11 Mare Basalt 10020	86
	9.3	Intensity of Recorded Fields from NRM	88
	9.4	Paleomagnetism of Melt Breccias	96
	9.5	Paleointensity Summary	100
	9.6	Crustal Magnetism: Magnetic Anomalies	100
	9.7	Summary	102
		References	102

10	**Lunar Magnetism in the Grand Scheme of Lunar History**	103
	10.1 The Kona Conference and a Model of the Moon's Origin	103
	10.2 Models of the Lunar Magnetism	107
	10.3 Summary	109
	References	110

Appendix A: Rotational Mechanics Background 111

Appendix B: Units 115

Appendix C: The Alphabet Soup of Remanent Magnetisms 117

Appendix D: Magnetometers for Sample Measurement 119

Appendix E: The Generation of Magnetic Fields in Electrically Conducting Media 123

Appendix F: Theory of TRM 127

Epilogue 129

Introduction

Of all the millions, who have gazed at our beautiful moon over the millennia, those of us lucky enough to be alive in 1969, were the first to be able to witness human beings on the moon. Those grainy TV pictures are forever in our memories. Little did I know that the samples that Armstrong and Aldrin were collecting would play such a role in my life. At that time, my research group and I were busy trying to understand what happened when the geomagnetic field reversed its polarity and other aspects of the magnetic record of terrestrial samples. That was plenty to keep us busy.

When Armstrong, Aldrin and Collins returned safely to earth, NASA made their precious cargo of samples available to the world's scientists. It was then possible to start to answer the fundamental questions about our moon: What is it made of, how was it formed, how old is it, what was the difference between the dark Mare and the lighter regions, were those circular features volcanoes, or impact craters? Was the moon an evolved body like the earth with a crust, mantle and core? Its origin was totally mysterious. Capture was very unlikely because it was known that the probability of the necessary combination of orbits was so low. Trying to spin it off the earth faced other major difficulties. Evidently, we had lots to learn.

I was lucky enough to be involved in studies of lunar magnetism from the earliest Apollo days because Professor Nagata used to work in our laboratory in the University of Pittsburgh, so that when he was awarded some of the first samples to be studied, we helped with the measurements. In addition, I also had the good fortune to be on the lunar sample planning committee. This put me at the center of discussions of the early work on the samples and to hear the experts in other fields first hand and informally.

In the near half-century following the return of the samples, we have gone a long way to answering many of those fundamental questions. As is so often the case, the new work also raised more questions. One of the puzzles from the start was lunar magnetism, which still remains somewhat problematic, but may finally be reaching some resolution. Magnetic measurements on the lunar surface, from satellites in orbit around the moon, and on the returned samples all revealed this mysterious magnetization. Yet the moon has no planetary magnetic field actively generated as does earth. When rocks like the lavas of the Hawaiian Islands cool, they acquire what we call a paleomagnetic record of the geomagnetic field. A remanent magnetization is stored in the rock from which we can recover the direction and strength of the field in which they cooled. On earth this field arises from the dynamo, in the fluid outer core of the earth. Was that what happened on the moon? Did the moon have a molten core giving a lunar dynamo field, or was lunar magnetism more exotic, perhaps related somehow to giant impacts on the moon?

This little book does not aspire to be a definitive scholarly text, but rather to tell, in as simple and entertaining a style as possible, the story of the great adventure of Apollo, of our new understanding of our moon, of the puzzle of lunar magnetism and of the fun of trying to find out what happened. As you will see this story is far from over for lunar magnetism, but we are making progress.

I have adopted an historical approach throughout, tracing the development of our ideas of the Earth–moon system, of the birth of the space age, of paleomagnetism and finally of lunar magnetism and its evidence for an early lunar dynamo. It is my hope that the book can be enjoyed by non-scientists, who like science and history, and by scientists not familiar with lunar science, if they skip some of the introductory material. In a further effort to serve these two masters, I have included in the notes for each chapter a few key references to the lunar scientific literature.

Before I begin the story, I would like to acknowledge the contributions of my colleagues in our research group over the years. It is one of the great joys of academic life to lead a research group. One is surrounded by bright young people to teach and from whom one constantly learns. In my group, first at the University of Pittsburgh and later at the University of California at Santa Barbara, I had the good fortune to work with outstanding colleagues. In particular, in the lunar efforts Stan Cisowski's contribution was pivotal and was the foundation of our work in the Apollo days. I also had equally outstanding senior colleagues in Bob Dunn and Vic Schmidt. Vic sadly died at far too young an age, but Bob continues to work with me nearly 50 years on.

Finally, to have worked on the lunar samples is a great honor. One only goes to the moon for the first time once! Let us hope our efforts are worthy of the opportunity. My thanks go to the many, many people, who made it all possible.

Chapter 1
The Moon in Antiquity and in the Development of Modern Science

To begin our journey to understand the puzzle of lunar magnetism, we first look at the long history of the development of our understanding of the earth moon system to set the stage for Apollo.

1.1 The Moon in the Greek World

Not surprisingly the moon plays a role in the mythology of ancient peoples. From the earliest civilizations in the near east onwards, all seem to have had lunar deities. The Greco-Roman tradition establishes female dominance in Western

Fig. 1.1 Selene visits Endymion in his eternal slumber

M. Fuller, *Our Beautiful Moon and its Mysterious Magnetism*,
SpringerBriefs in Earth Sciences, DOI: 10.1007/978-3-319-00278-1_1,
© The Author(s) 2014

Civilization lunar mythology. Selene, whose name is remembered in selenography, as studies of the moon are sometimes called, was a Greek goddess of the moon. She was a daughter of the Titans, was in love with Endymion, and asked her father Zeus to grant this beautiful mortal immortality. Zeus acquiesced, and Endymion was visited by Selene in his eternal slumber (Fig. 1.1).[1] Together, the pair had 50 daughters. Phoebe, another lunar goddess, was of the first generation born of the Titan deities of the Golden age. Her consort was Coeus her brother, who sired Leto and Asteria. The former mated with Zeus and bore Artemis (still another moon goddess) and Apollo. Let us leave this splendidly salacious world of Greek mythology and go to the first naturalist philosophers, who tried to understand the moon in a modern sense.

In turning to the earliest science of the moon in the west, it is natural to stay with the Greeks, but first let us try to imagine ourselves in their world about 2500 years ago. We are in Argos one of the many city states of ancient Greece. We are proud of the famous statues and the art of our city. We have history that goes back hundreds of years to the ancient glories of Homeric heroes. We are standing on the old land of these heroes, which is about as solid and unmovable as anything can be. In the day, we watch the sun rise in the east cross the sky and set in the west, followed at night by the moon, and all the stars of the night sky. We know the moon has phases (Fig. 1.2),[2] unlike the sun, or any of the other heavenly bodies, as far as we can see. The moon seems to keep time for us with its regular sequence of monthly changes. It follows the same procession across the night sky. Is it not natural to think the whole sky and everything in it rotates about the earth. This was the starting point for the earliest Greek philosopher naturalists, who sought to understand the world.

The first Greek philosopher we meet is Thales of Miletus ($\sim$ 620–546 BC), the first of the seven sages of ancient Greece named by Plato. What we know of his work was from others, for no writings of his have come to us. He appears to have been the first to ascribe natural causes to phenomena, rather than accepting them as the whims of the Gods. Recognizing the importance of Egyptian learning in his

---

[1] Selene (Diana) visits Endymion in his eternal slumber, Painted by Sir Edward Poynter (1836–1919), Victorian portrait and historical painter, President of the Royal Academy, educated Brighton College and Oxford University. This is one of several Victorian era paintings of the subject. Endymion was also immortalized in the poem by Keats with the opening line of the first stanza—"A thing of beauty is a joy for ever …" Some of the many books that may appeal: *The Library of Greek Mythology* (Oxford World's Classics), Apollodorus Translated by Robin Hard, 1997, Oxford University Press. This is the only mythology text to survive from the classical era. How much of it was written, or edited by Apollodorus is not clear. It covers myths from creation to the Trojan War; *The Complete World of Greek Mythology*, Richard Buxton, 2004, Thames and Hudson, this book gives helpful chronologies; *The Greek Myths*, Robert Graves, Penguin Books, 1992, First published in two volumes by Pelican books, 1955.

[2] Compiled with inclusions from various NASA figures.

1.1 The Moon in the Greek World

Fig. 1.2 The phases of the Moon—a modern version

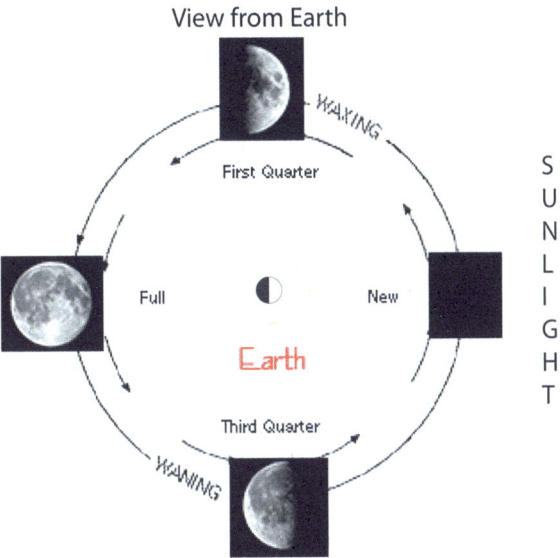

time, Thales traveled there and brought geometry back to Greece. He knew that the earth was round and that the moon shone by reflected light from the sun. He realized that a solar eclipse took place when the moon passed in front of the face of the sun. He was also the first Greek to predict an eclipse successfully. How he made the prediction is not clear, but he may well have made use of the extensive Near Eastern records. One effect of the eclipse was remarkable; the Medes and Lydians had been fighting indecisively for about 5 years, but with day turning to night during a battle they decided it was time to try diplomacy.

From our viewpoint, his work on the orbits and diameters of the sun and moon is particularly interesting. Given the geocentric system with which Thales worked, the fraction of the sun's orbit corresponding to the sun's diameter, yielded the ratio of the sun's diameter to the length of its orbit around the earth. Here Thales was advocating that the sun was much larger than was generally thought at the time. He estimated the fraction of the solar orbit represented by the diameter of the sun and showed that given any reasonable orbit for the sun and its diameter must be very large. For the moon a similar ratio emerged, so a comparison of the two could be made. It was long way from solving the problem, but it was a start. For the solution to these problems of the size of the moon and its orbit, we will have to wait roughly three centuries.

Thales seems to have been aware of the power of certain iron ores from Magnesia to attract particles of iron—

> The magnet's name the observing Grecians drew
> From the magnetic region where it grew ...,

as Lucretius wrote. At the very outset of our story, we find interest in magnetism.

An often told story of Thales is that on one occasion, when he was paying more attention to the heavens than to where he was walking, he fell into a ditch. As the story goes, a young Thracian lady suggested to him that he should pay more attention to the earth than to the heavens, but Thales was vindicated in a manner, when his studies of the weather suggested a good forecast in olive futures and he became rich.

Of all the ancient Greek philosopher naturalists, it was Aristotle (384–322 BC), whose ideas eventually dominated the "scientific" history of the west in mediaeval times. He was a student of Plato at the Academy, but more of an empiricist than Plato. For example, he did not accept Plato's theory of forms. He thought that properties were intrinsic to an object, rather than abstract universals independent of the objects. From a scientific point of view, his emphasis on observations is central. He was the tutor of Alexander the Great and questions arose concerning how consistent Aristotle was to his written ideas of the polis, in this service to Alexander. He had many enemies in the Athens when he returned. Eventually he fled from Athens to Chalcis to avoid public trial and died there within a year. His texts were preserved and studied in the near east and were known in more detail there, than in Europe. Some of this knowledge in all probability came back to Europe with the Moors in Spain. Certainly, his ideas and in particular his picture of the universe became dominant in the western civilization during the middle ages. For him, it was a geocentric system with moon, planets and sun all orbiting the earth in perfect circles and beyond the fixed stars lay the prime mover. Let us leave to the philosophers, who or what the unmoved prime mover of the system is, and simply accept that in Aristotle's model the motion was transferred inwards from the prime mover.

With time, the center of Greek astronomy moved to Alexandria and there, in the 3rd century BC Aristarchus (320–250 BC) proposed a heliocentric model of the solar system. It was not well received and had to wait more than a thousand years to be accepted. Its rejection appears to have been based upon two arguments. First, common sense told one that the earth was not moving. Second, if the earth were moving, one should see the parallax in the stars. This is the same effect one sees when traveling in a train or car, whereby nearer objects move with respect to the distant background. If the earth is moving, one should see nearby stars move with respect to the more distant background stars. They do not, so therefore the earth is not moving, QED, as we used to write on proofs in high school geometry. The problem was that it was not until 1838 that telescopes had sufficient resolution to detect this parallax.

Another truly remarkable achievement of Aristarchus was to solve the problem Thales had begun to investigate, namely the relative distances of the sun and moon from earth. He assumed that when the moon was at the true first quarter (Fig. 1.2), it forms a right triangle with the earth and sun (MES in Fig. 1.3).[3] He then noticed

---

[3] Fisher, C., 1943, The story of the moon. Doubleday, Doran and Company. This book provides an excellent history of the development of ideas of the moon including the determination of its size and orbit. It also has an extensive discussion of the origin of the craters on earth and on the moon, which we will use later. He credits Sir Richard A. Proctor with the first suggestion that they were impact craters in 1873.

1.1 The Moon in the Greek World

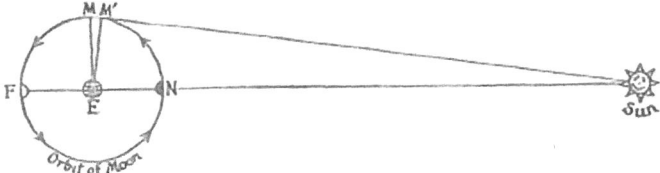

Fig. 1.3 Aristarchus measures the relative distances of moon and sun from earth

that when we view first quarter from earth, the moon is actually not quite at the true first quarter position, but is at M'. It would therefore take less time from the new moon to the observed first quarter than from the observed first quarter to the full moon. The differences he measured were badly in error and the angle MEM' turns out to be far smaller than he could have obtained from his observations, but the method is good. The times he obtained for the difference between the new moon and the true first quarter and the new moon and the observed first quarter gave him, the angular fraction of the moon's period represented by MM'. This was the angle MEM'. Turning to the right angle triangle M'SE, the angle M'ES follows, and then M'SE the third angle of the triangle. We would now use trigonometry tables to say that, the sine of the angle M'SE gives moon's distance from earth divided by the sun's distance from earth. However, this convenient approach was not available to Aristarchus and he had to use a more complicated method to get the same result.

The third remarkable achievement of Aristarchus was to use a lunar eclipse to measure the relative sizes of the earth and moon from the time it took the moon to pass through the earth's shadow, but we will leave this to the completion of the problem provided by Hipparchus that gives the absolute values.

In the next century Hipparchus ($\sim$190–120 BC) championed the need for accurate observations. He was an early master of trigonometry. Unfortunately, again nothing has come down to us of his writing. However, both Ptolemy and Pappus of Alexandria describe his method of measuring the distance to the moon from observations of a solar eclipse from two sites on earth. When a total eclipse was observed at Syene, at Alexandria 1/5 of the sun was still visible. The angular size of the part of the Sun visible from Alexandria was therefore 0.1°, that is 1/5 of the angle subtended by the sun's disc, which is 30 arc seconds or 1/2°. Thus, he knew the distance between Alexandria and Syene and the angle subtended by that distance at the moon. From this he could get the distance to the moon and the value he got was 250,000 miles which an excellent approximation. I will leave it to the reader to have fun figuring out how he might have done this (Fig. 1.4a).

Let's now see how he might have used a lunar eclipse to find the size of the moon (Fig. 1.4b). This can only occur when the plane of the moon's orbit intersects the ecliptic plane and at full moon. The method relies on the size of Earth's shadow at the orbit of the moon during the eclipse, which depends primarily on the size of the Earth. Hipparchus would have known from Eratosthenes ($\sim$276–194 BC) the size of Earth. By timing the passage of the moon through Earth's shadow

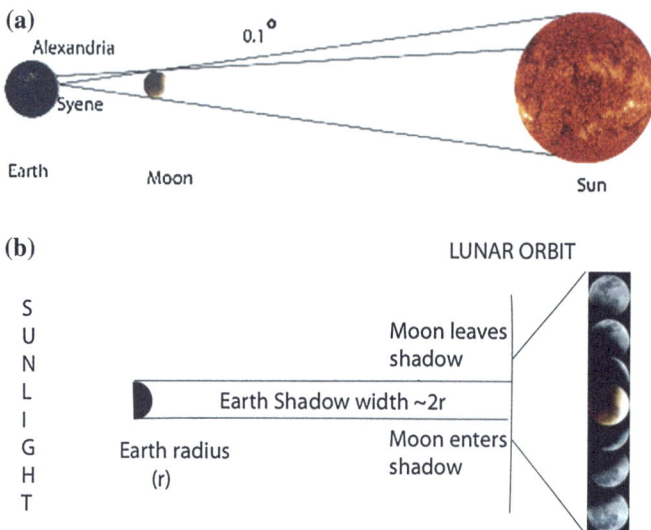

Fig. 1.4 Simplified diagrams of determination of (**a**) distance to the moon from a solar eclipse and (**b**) size of the moon from a lunar eclipse

he might have argued as follows. The moon moves around Earth in a circle and it takes one month to do so in its period (T). With the radius of the moon's orbit R, the distance to complete the orbit is $2\pi R$. With the radius of the earth of r, the shadow's width at the moon is close to 2r. Assuming that the moon moves around Earth at constant speed then and takes time t to cross earth's shadow, then

$$2\pi R/2r = T/t,$$

which gave a value of R/r of 60. Using just the radius of the earth from Eratosthenes, Hipparchus would have come close to the present value, an astonishing achievement of the Greeks. Aristarchus also used this method as we saw, but he lived before the determination of the size of Earth by Eratosthenes and so could not get the absolute values.

Hipparchus also compiled a star catalogue and observed the effect of the precession of the rotation axis of earth. As we now know, the earth's rotation axis is not fixed in a constant orientation with respect to the stars, which means that what we now call the north star is only transiently aligned with geographic north. His observations told him that in the geocentric scheme, the axis about which the heavens rotated around the earth changed with time. He achieved this remarkable feat by using the shadow of the earth on the moon during an eclipse of the moon, when the sun, moon and earth are in a straight line. He could then determine the point on the celestial sphere in the backdrop of the stars, which was opposite to that of the sun. When he compared his results with earlier observations he found

1.1 The Moon in the Greek World

Fig. 1.5 Retrograde motion of Mars—heliocentric interpretation

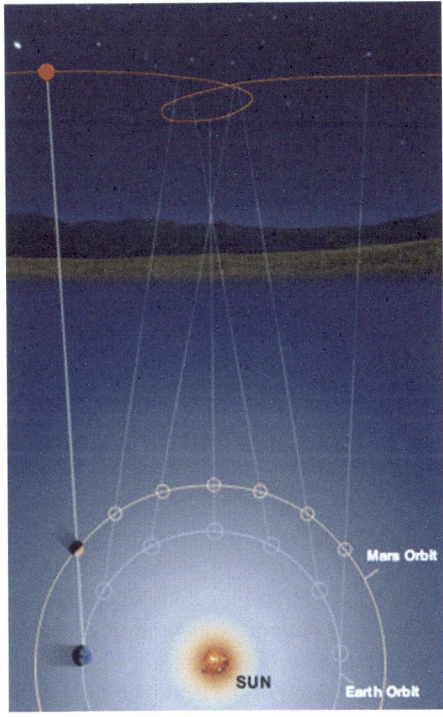

that over 169 years the intersection point had moved 2°. We will discuss a modern view of precession later.

Hipparchus developed a geocentric model of planetary motion, but hit a problem. If all the planets were moving around the earth in the same direction, why did they sometimes appear to be going forwards in their orbit and sometime going backwards (Fig. 1.5).[4] By then the Greeks had already recognized the retrograde motion of planets. However, he found that he could explain the observations by having the planets move in a relatively small circle (the epicycle), the center of which moved about the earth in a circular orbit (the deferent). The orbital motion of the center of the epicycle around the deferent and about the earth was always in the same direction.

The eventual result was the system of deferents and epicycles that was incorporated in the Ptolemaic model (Fig. 1.6). The planet moved anticlockwise in the epicycle, which itself moved anticlockwise around the earth along the deferent. Claudius Ptolemy (AD90–AD168) extended the idea of deferents and epicycles in his system to include epicycles on epicycles and the introduction of the equant about which the center of the epicycle moved. This work was carried out in Roman

---

[4] Courtesy NASA on line image.

Fig. 1.6 Ptolemaic system with epicycles

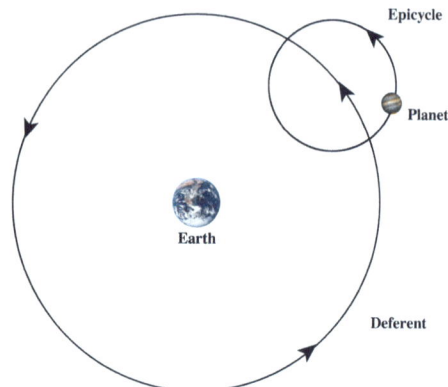

Alexandria. Ptolemy's Almagest dominated astronomy for more than a thousand years and Ptolemy was probably the greatest of the Greco-Roman astronomers after Hipparchus, although there have been counter views.[5] In addition to his astronomy, he was an accomplished mathematician and a geographer, who summarized the learning of his time in this field in his "Geographia". His treatise on Astrology—Tetrabiblos dominated that subject for more than a thousand years. It appears that Ptolemy was more interested in the accumulation of astrological data than its use to predict outcomes of particular events.

No discussion of the achievements of the Greeks concerning the moon would be complete without mention of the Antikythera mechanism, the remains of which were found in 1900 in a Roman shipwreck roughly half way between the southernmost Pelopennesus and Crete. It appears that the wreck took place sometime in the 1st century AD and the device constructed more than 100 years earlier. Some have suggested that the great naturalist mathematician Archimedes may have been involved in the development of the device, but he died well before its construction. However, the school Archimedes founded may indeed have played a role. After Herculean efforts, it has recently been analysed and shown to be a mechanical computer that predicted lunar and solar eclipses, as well as the motion of the planets and indeed the whole earth centered universe.[6] The sophistication of the mechanisms in giving the movement of the moon makes this one of the foremost wonders of the world of the Greeks.

Tragically many of the precious texts of antiquity were lost when the Library of Alexandria burned during Roman times. The next great center of learning to arise was Baghdad. This was not until the 8th century AD, when reconstruction of the city was underway to serve as the capital of the Islamic empire of the Abbasids.

---

[5] Newton, R.R., The crime of Claudius Ptolemy (1977). There are suggestions that Ptolemy may have falsified some of his data and the case is made most strongly in this book.

[6] Marchant, J., *Decoding the heavens: A 2000-Year-Old Computer- and the Century long Search to Discover its Secrets*, 2010, Heineman, In this book the incredible story of the interpretation of the mechanical secrets of the Antikythera mechanism is told. This topic was also addressed in a recent Nova program entitled *Ancient computer*.

1.1 The Moon in the Greek World

From its earliest time it became a center of learning and business, soon surpassing Ctesiphon, the older Persian capital. It was here in Baghdad that the ancient Greek texts were preserved and studied. It was also a major center of business, original mathematics and astronomy. Again precious records were lost in 1248, when Baghdad was sacked and most of the city's inhabitants massacred by the Mongols under Hulagu Khan, one of innumerable grandsons of Genghis Kahn. A little earlier in 1204 the crusaders of the Fourth Crusade were busy pillaging Constantinople, destroying untold numbers of ancient texts, so we should remember that our Christian forbearers were not guilt free in this era.

1.2 The Beginning of the Age of Modern Science

By the time the Middle Ages were drawing to a close, classical Greek texts were established in Europe. The Aristotelian ideas were accepted by the church and formed the basis of a Christian synthesis by Thomas Aquinas (1224–1274). It is ironic that the ideas of Aristotle, who had emphasized the importance of observations, became accepted dogma of mediaeval times that could not be questioned by new data, or ideas. The essential features of the synthesis were that (1) motion of the heavens was in perfect circles, (2) the heavenly objects were perfect and unchangeable and (3) the earth was at the center of the universe and everything orbits about it. As we have seen, improved observations by the Greeks had already required the use of deferents and epicycles to account for the retrograde motion of planets. Epicycles on epicycles were also required as formulated by Ptolemy in the Almagest (Fig. 1.6) and the model had become increasingly complicated. Thus, the stage was set for the great struggle between dogma and empirical science.

With the work of Nicolas Copernicus (1473–1543), the view of the earth as the center of the universe, with everything in orbit around it was replaced with a simpler model, in which the earth and the planets moved around the sun. Giordano Bruno (1548–1600) took these ideas further and saw the sun as a star and suggested that other "solar systems" around other stars existed and that some were likely inhabited by intelligent life. These ideas were determined to be heretical by the Roman Catholic Inquisition, and Bruno was brought to trial, convicted, and burnt at stake. Now 500 years on, astronomers are showing that he was right about other "solar systems". We do not yet know whether his second idea was correct. Yet, it seems a pretty safe bet, even though we do not know how life started on earth. One suggestion is that life may have originated somehow from the organic molecules brought to earth by comets, asteroids and meteorites, after the early catastrophic bombardment had calmed down. In this case, as Bruno suggested, a similar process may have taken place in "solar systems" all over the universe. This is perhaps taking us too far afield from our moon, so let's return to our story.

Tycho Brahe (1546–1601), like some of his Greek antecedents, had become convinced that accurate observations were the key to any improvement in astronomy. New possibilities were opening up in instrumentation and Brahe was

Fig. 1.7 Kepler's 2nd law of planetary motion

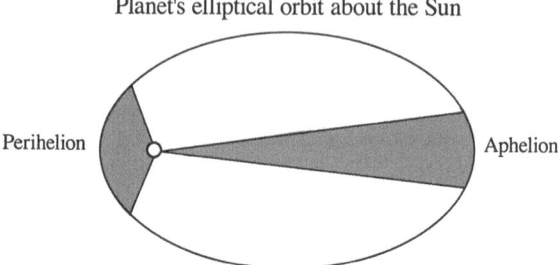

given a small island by Frederick II of Denmark to build an astronomical observatory. With his own instrumentation, which he constantly checked and recalibrated, he worked for some 20 years observing and training young astronomers. His observations were the best ever made before the introduction of telescopes, probably reaching the limit of accuracy using the unaided human eye. After falling from favor with Danish royalty he left his island and eventually lived in Prague. His observations were not published in his lifetime and his instruments were lost, but fortunately his assistant Johannes Kepler (1571–1630), who had been hired to calculate planetary orbits, used Brahe's data brilliantly to get the three fundamental laws of planetary motion: (1) planets move round the sun in ellipses, with the sun at one focus of the ellipse, (2) the line connecting the Sun to a planet (the radius vector) sweeps out equal areas in equal times and (3) the square of the orbital period of the planet is proportional to the cube of the mean distance of the planet from the sun. We can see that the second law reflects the variation of the speed of a planet in its elliptical orbit—it moves slowly when it is more distant from the sun, and speeds up as it falls in towards the sun from its aphelion to its perihelion (Fig. 1.7). These laws had to await the genius of Newton for their interpretation and explanation. Before that we need to recognize the contribution of the other genius, who ushered in the modern world of science.

Brahe (1546–1601), Kepler (1571–1630), and Galileo (1564–1642) were all contemporaries. Unlike Bruno, Galileo was not killed for his ideas, but he was placed under house arrest for his support of the Copernican model. As we shall see, although this was clearly an injustice to Galileo, it may have been fortunate for us because during this time he wrote up much of his earlier work. His views had been found to be "vehemently suspect of heresy". From the point of view of the clash between dogma and empirical science, his development and use of the telescope to see the lunar surface was crucial.[7] He saw that the moon was not a perfect sphere of some celestial material, but had mountains and valleys like earth (Fig. 1.8). This was certainly not compatible with the Aristotelian model and was heretical.

Again Galileo had seen that the moons of Jupiter moved around Jupiter, so that astronomical bodies could orbit other bodies than earth. He had also seen the

---

[7] On line image in public domain but see Galileo, *Sidereus Nuncius* (The Starry Messenger) (1610).

1.2 The Beginning of the Age of Modern Science

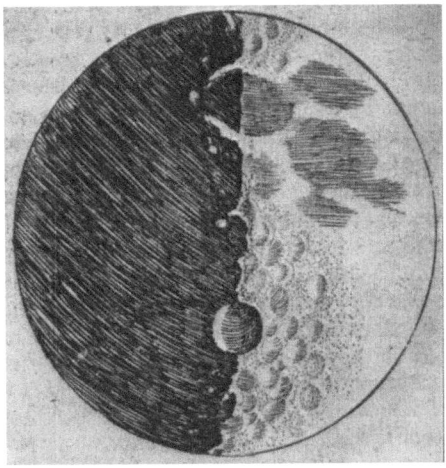

Fig. 1.8 Galileo's map of the moon

phases of Venus and understood their significance. Galileo was a man of faith and one can see the tightrope he cleverly walked from the following quote "I think in the first place that it is very pious to say and prudent to affirm that the Holy Bible can never speak untruth—whenever its true meaning is understood."

It is worth pausing to think about the leap to the modern world that came with Galileo and Newton. In this book, we will often be humbled as we contemplate the contributions of genius, but in an important sense these two were the giants that brought us the methods of modern science. Galileo wanted to test whether bodies fell at the same speed independently of their mass, not to argue about it, or to consult authority. He may have tested it using the leaning tower of Pisa, but whether that particular test was carried out or not, the key for him was to test an idea properly. Late in his life under house arrest, Galileo used the inclined plane method of observing "falling" bodies,[8] an experiment that I think that I can still dimly remember repeating in my high school physics lab. He allowed different balls to roll down a trackway on a gently inclined slope, so that he could time the distance they rolled and measure their velocity. He found that in equal times they all rolled a constant pattern of distances, no matter what they were. The longer the ball rolled the greater was the velocity. This was in a simple ratio to time determined by a constant. The distance that the ball rolled was related to time in a more complicated way. He showed that the distance the ball rolled is proportional to the square of the time involved, Galileo had managed to test his ideas on falling bodies and establish the relevant mathematical laws and reinforced his recognition that the key to understanding the world was through mathematics. With Newton more complicated phenomena were analyzed through simplified models that could be tested mathematically against data. This is the bedrock of modern science. He

---

[8] There was a very good discussion of this work of Galileo in a Nova program entitled *Galileo's Experiments*.

could then show that Keplerian orbits required an inverse square law for the attractive force of gravity. The moon's orbit is determined by the balance between this gravitational force, pulling it towards the earth, and the moon's inertia maintaining its motion unchanged, or if you prefer, the balance between its centrifugal force and the earth's gravitation attraction. The orbit is nearly circular and at a distance of ~240,000 miles (384,000 km) from earth. The moon travels at a speed of ~2,300 mph (3,700 kph) about the earth. To be more accurate both the earth and moon rotate about the barycenter (center of mass) of the earth moon system, which is within the earth.

1.3 Evolution of the Earth Moon System: Tides and Precession

There are two additional topics we need to understand for our story. One is the tide and its effect on the history of the earth moon system and the second the precession of the earths rotation axis. Appendix 1 provides a little more discussion of energy and angular momentum, which may be helpful in this next section for the non-scientist.

The earth's rotation carries the tidal bulge on the earth forward from the point immediately beneath the moon to a point where there is a balance between the frictional force and the tide raising force (Fig. 1.9). Friction opposes the tide raising force and has a number of sources, including friction between the waters of the oceans and the ocean bottom. The gravitational attraction of the bulge on the moon speeds up the moon's motion in its orbit. In turn the moon exerts a force back on the earth to slow its rotation. The result is that the angular momentum of the system is conserved and the moon speeds up in its orbit and recedes, while the earth rotates more slowly as time passes. This was first suggested by Edmund Halley. The recession rate has been measured directly by timing laser light pulses going from Earth to Moon and being reflected back to Earth by reflectors left on the moon by Apollos 11, 14 and 15. Geological evidence of the long history of this effect has come from sediments that are laid down in estuaries with large tidal flows. From these sediments daily, monthly, and annual records can be obtained. The results give a day of 21.9+/−0.4 h and 400+/−7 days in the year ~620 million years ago.

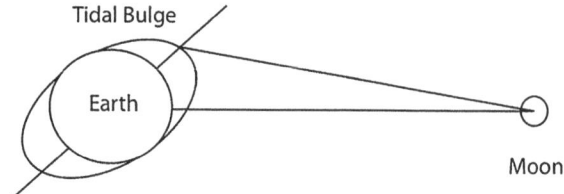

Fig. 1.9 The tides and evolution of earth moon system

1.3 Evolution of the Earth Moon System: Tides and Precession

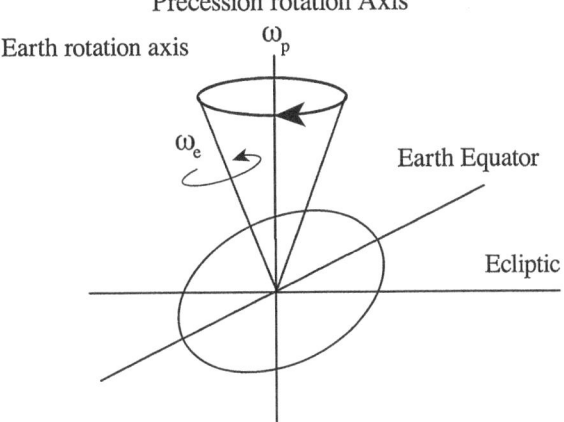

Fig. 1.10 Precession of the rotation axis of the earth. ω_p is the precessional angular velocity and ω_e is the angular rotational velocity of the earth

The moon may play another role for the earth and life in stabilizing the rotation axis, so that the angle between the perpendicular to the ecliptic plane in which the planets lie and the rotation axis is relatively constant. The rotation axis precesses about the perpendicular to the plane of the ecliptic as Hipparchus discovered 2000 years ago. The effect is illustrated in Fig. 1.10. The axial tilt, or obliquity is 23.5° at present. It is of course this axial tilt that give us the seasons with the northern summer, when the north pole is inclined towards the sun and southern summer when it is the southern pole that points towards the sun. The inclination of the axis to the ecliptic varies with an approximately 41,000 year period. The direction of precession is opposite to the earth's rotation. The period of the earth's precession, the time taken for the earth's rotation axis to complete its cycle and return to pointing at the same star in the night sky, is $\sim 26,000$ years. The precession is caused by the gravitational attraction between the sun and moon on the equatorial bulge of the earth. The lunar contribution is the greater by a factor of about 2 (Fig. 1.10).

The pull on the equatorial bulge to bring it into alignment with the ecliptic has an analogous effect to the weight of a spinning top pulling its center of mass downward, when the top departs from the vertical. When you push the rotation axis of a spinning top from the vertical, as you will perhaps remember, the spin axis of the top will respond by moving in a circle about the vertical. The top is then precessing, as does the earth's rotation axis.

The role that the moon has in stabilizing the rotation axis of the earth and precluding massive changes in the obliquity is a key to the welfare of life on earth, but there are also significant contributions from other planets especially Jupiter.

1.4 Summary

With this rapid scan of the millennia, we have seen the history of methods of determination of the size of the moon with its diameter of 2,159 miles (3,474 km), and its nearly circular orbit at a distance of $\sim 240,000$ miles (384,000 km) from earth. The moon travels at a speed of $\sim 2,300$ mph (3,700 kph) about the earth. We saw the introduction and final acceptance of a heliocentric model, the establishment of the scale of the solar system with the orbits of the planets about the sun, the understanding of the tides and of their effect on the earth moon system, and the precession of the earth's rotation axis. We are now ready to begin our look at the modern studies of the moon and the earth in the Apollo era. As we shall see much later, precession and the moon's recession may be key points in explaining the history of lunar magnetism.

Chapter 2
Lunar and Earth Sciences at the Time of the Apollo Landings

We will now focus in on the moon itself in particular on the origin of its surface features. If they are indeed giant impact craters they might account for lunar magnetism. We will also introduce the topic of paleomagnetism some other relevant ideas from mid 20th century geophysics and geochemistry which are key to our story.

2.1 Early Modern Studies of the Moon and Impact Cratering

If we are looking for a founding father of modern lunar science, a good candidate is Ralph Baldwin (1912–2010). He was, a gentleman scientist, representative of a past era, with a full time job as a Vice-President of the family business and did his science because he loved it. He had however been well trained as an astronomer. He was no amateur, as far as his knowledge and skills were concerned. In the 1940s he became interested in the moon and the origin of the features on its surface. The current view was that they were of volcanic origin, but one gets the impression that the moon was very much on the back burner of the science of the day. Astronomers were not too much concerned with such local objects as the moon and planets, and geologists had not yet become very interested in them.

Ralph Baldwin's "The Face of the Moon" published in 1949[1] may not have attracted much attention then, but it is a beautiful book. In the preface he lays out its structure, which leads after a historical introduction to his observations and to his thesis, which is that,

---

[1] Baldwin (1949). This book covers a mass of interesting material with much more detailed treatment of work in the 19th and 20th centuries than given here. Another source, which may prove useful is "Shock Waves & Man" by I.I.Glass. It covers the broad field of shock effects in a similar style to that I have aspired to in this book.

The single process which offers promise of being the one which actually operated to build the moon's surface structures is that of meteorite impact. Meteorites travel very fast and have great kinetic energy. When one strikes the moon, this energy must be released suddenly and thus an explosion occurs. If the meteorite is large enough and moving fast enough, it will carry sufficient energy to produce the equal of any crater now found on the moon.

The book presents detailed data on craters, comparing for example features of bomb craters and terrestrial impact craters with lunar craters. He also emphasizes that in comparison with scale of volcanic activity and the calderas seen on earth, the lunar phenomena is far grander and could not possibly be caused by volcanic activity. In this, he was following in the footsteps of the geologist Grove Karl Gilbert (1843–1918), who had come to the same conclusion.

The history of the recognition of impact craters on earth followed a somewhat similar course. In the 1920s Walter Bucher studied a number of features, which he called cryptovolcanic.[2] He realized that they must have originated in a massive explosive event, but he thought that it was internal, or volcanic in origin. The geologists John D. Boon and Claude C. Albritton, visited Bucher's sites but interpreted them as impact craters.[3]

Bob Dietz (1946) also argued for an external origin for the features that Bucher had described and gave them the name astroblemes. Long before this, at least one crater on earth had been recognized as an impact feature. This was meteor Crater in Arizona, which is about 50,000 years old (Fig. 2.1). In 1903, Daniel Barringer identified this as an impact crater. Pretty conclusive evidence had already been seen by the first Europeans to pass by, when they saw tons of oxidized iron meteorite fragments in the surrounding area. Barringer had thought that he would be able to recover large amounts of very valuable iron from the crater itself, but had not realized that most of the meteorite had in fact been vaporized, as Boon and Albriton had previously suggested.

A key player in the early modern studies of terrestrial impacts was Eugene Shoemaker and in Chao and Shoemaker (1960) demonstrated the presence of coesite in Meteor Crater. Coesite is a form of silica that could only be formed by the level of shock associated with impacts, or nuclear explosions. This then was the critical observation, the test, which proved the extraterrestrial origin of the feature. No volcanic event could provide the necessary shock level to generate this mineral.

At this time, the Moon was emerging from astronomical semi-obscurity "to claim renewed interest and attention".[4] Lunar observations and theoretical discussions were coming to the fore and the moon was ceasing to be on the back

---

[2] Bucher (1933). I remember reading this book as an undergraduate, but was not smart enough to see that the scale of meteoritic impact was so much greater than volcanic events of the type that Bucher envisioned.

[3] Boon and Albritton (1936). This paper was presented at a meeting in Chicago, but the authors do not seem to have received much credit for their idea in the recent literature.

[4] Kopal (1962). This book is a convenient starting point for a picture of the state of lunar science at the beginning of the 1960s.

2.1 Early Modern Studies of the Moon and Impact Cratering

Fig. 2.1 Meteor crater Arizona

burner. A common view expressed by Harold Urey was that "The moon accumulated independently of the Earth and represents a more primitive object than the terrestrial planets". Following this line of thought he assumes that the moon was captured. However, he was quick to recognize that these ideas may be changed by direct exploration. By 1960, the Russians had already landed Luna 2 on the surface of the moon and Kopal notes it "will soon be followed by others carrying instruments and eventually men to the moon".

2.2 Geology and Geophysics in the 1960s

We also need to remember how the geological and geophysical sciences stood in the 1960s, for they were to be critically relevant to lunar exploration. It was a time of great turmoil with the birth pangs of plate tectonics and the growing introduction into geology of the methods of modern physics and chemistry.

The queen of the geophysical sciences is seismology, the study of earthquakes and the analysis of the passage of seismic waves through the earth to yield its internal structure. In about 150 years, we have learnt exquisite details about our earth from seismology. This is not to say that there was no seismology before then. There was. The Chinese had an operational seismograph millennia ago, but modern seismology of the past 150 years has given us the key results that allowed us to determine the internal structure of the earth. In the early years of the last century sufficient seismograms were available to record the arrival times of seismic waves from earthquakes on a worldwide basis. At this time, R. Oldham recognized the different types of seismic waves that propagated away from the earthquake source: (1) P-waves, which are pressure waves, analogous to sound waves, with particle motion parallel to the velocity direction, (Fig. 2.2a), (2) S-waves, which are shear waves with their particle motion perpendicular to direction of motion (Fig. 2.2b), are unable to pass through liquid and (3) Surface waves—the waves which propagate along the earth's surface and do most of the damage in earthquakes.

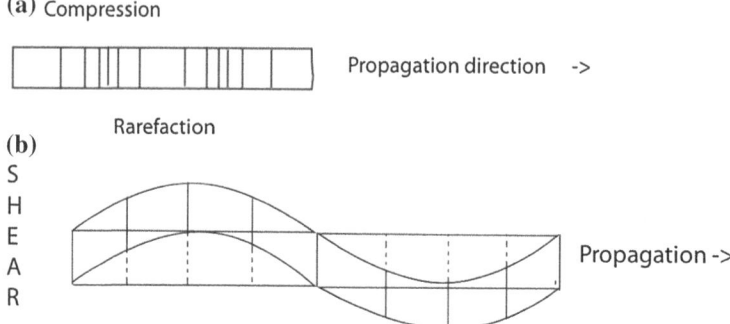

Fig. 2.2 Particle motion (**a**) in P-wave is parallel to the propagation direction giving regions of compression and rarefaction (**b**) in S-waves particle motion is perpendicular to the propagation direction

When it was found that S-waves disappeared at some angular distance on the earth's surface from the earthquake source, Oldham recognized that this must be due to the presence of a liquid core through which S-waves could not pass. Assume an earthquake at the north pole in Fig. 2.3,[5] and consider only the simplest trajectories of P and S-waves that travel southward in the mantle. Both P and S-waves will curve to shallower paths as the velocities increase as they go deeper into the mantle. The P-waves whose trajectory takes them into the fluid outer core will be refracted at the core mantle boundary (CMB) to take up a deeper trajectory because of the decrease in P-wave velocity in the outer core compared with the mantle. When the P-wave leaves the fluid outer core and reenters the mantle it will be refracted again, as is shown in the figure, so that there will be a gap in which P-waves are not seen—a shadow zone. The S-waves that stay in the mantle will follow a similar path to the P-waves, but no S-waves can enter the fluid outer core, so S-waves will not be seen beyond the grazing path similar to where the P-wave shadow starts. The last part of the deep earth structure puzzle was solved by Inge Lehmann in 1936, when she discovered the solid inner core.

The crust on which we live is separated from the mantle by the Moho, which is roughly 5–10 km down under the oceans and 35–90 km under the continents. The Moho is named for Mohorovicic the Croatian, who discovered it. He found that there were double sets of arrivals of P and S waves in shallow earthquakes, one of which had propagated entirely in the crust and the other had entered the upper mantle before returning to be observed at the surface. The boundary separates regions of P-wave velocities of 6 km/sec in the crust and 7 km/sec in the mantle. There are outcrops of the mantle rocks in ophiolites, which are sections of ocean floor that have been thrust up to the surface by tectonics, so we have seen the nature of the upper mantle under the oceans.

---

[5] Combination of NASA images Cosmos.

2.2 Geology and Geophysics in the 1960s

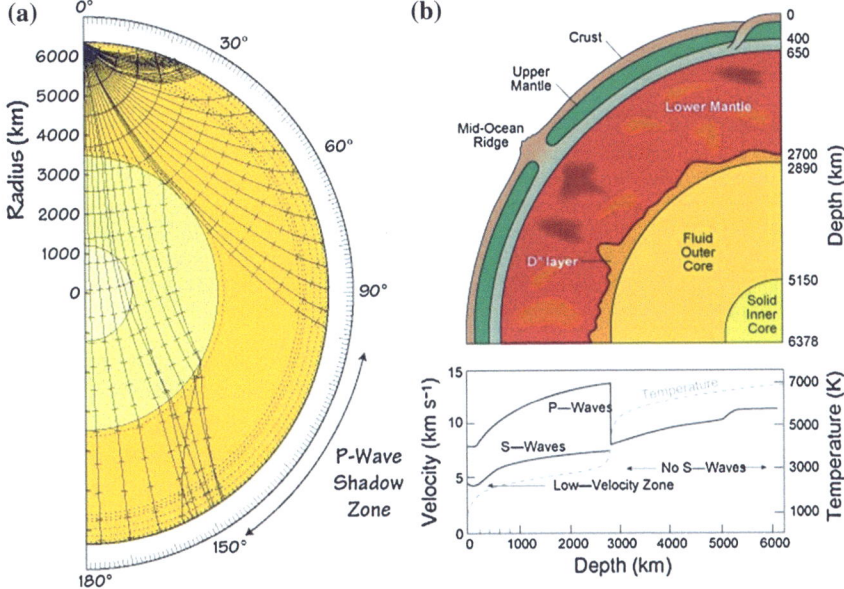

Fig. 2.3 Internal structure of the earth (**a**) P-waves and P-wave shadow zone and (**b**) summary figure

Radioactive age determination had matured and produced key results by the 1960s, such as the age of the earth of at 4.55 Ga by Claire Patterson and of 4.95 Ga for the solar system by John Reynolds. This method of radioactive age determination would be key in establishing the history of the moon. Rapid progress had followed Rutherford's initial work. The method depended upon knowing the decay rate of particular atomic species and measuring parent and daughter masses in a rock considered it as a closed system, from which nothing escaped and into which nothing went. From the remaining mass of the parent, the generated mass of the daughter product and knowing the decay rate of parent to daughter one can determine how long the system has been closed, although as usual it turns out to be not quite as simple as this might seem.

From Rutherford comes a classic anecdote. It concerns a formal presentation of his work, which happened to be in the presence of Lord Kelvin. That great man was famous for his opposition to an age of the earth that gave the long periods of time required by geologists and biologists. He maintained eventually from his cooling model that the age of earth could not be more than 20 million years. This was nowhere near enough time for Darwin, whose theory of evolution required at a minimum a far longer earth history, or indeed for contemporary geologists, who needed far more time for geology to work its ways. Lord Kelvin was not very charitable to his geological and biological colleagues calling their methods of determining the passage of time from phenomena such as the erosion rates of chalk absurd. However, while these geological estimates of the passage of time were certainly not very accurate, they turned out to be better than the ages from Lord Kelvin's method with its impeccable mathematics, based upon an incorrect model for the cooling of the earth. To return to our anecdote Rutherford said,

> I came into the room, which was half dark, and presently spotted Lord Kelvin in the audience and realized that I was in trouble at the last part of my speech dealing with the age of the earth, where my views conflicted with his. To my relief, Kelvin fell fast asleep, but as I came to the important point, I saw the old bird sit up, open an eye, and cock a baleful glance at me! Then a sudden inspiration came, and I said, 'Lord Kelvin had limited the age of the earth, provided no new source was discovered. That prophetic utterance refers to what we are now considering tonight, radium! Behold! The old boy beamed upon me (Rutherford 1904).

As usual there are many lesser known heroes in these studies, who made key contributions. One such is Boltwood, who traced sequences of decay products including the uranium lead decay series, which is the basis of so much radioactive age determination. A geologist whose contributions have perhaps been underestimated is Arthur Holmes. He initially did major work in radioactivity and established the first quantitative geological time scale. He also suggested that convection in the mantle provided a mechanism to explain Wegener's theory of continental drift. His text "Principles of Physical Geology" was very popular among physicists coming into geology to test continental drift with paleomagnetism (Holmes 1944). It gave them a quick, but thorough introduction to modern geology.

The relatively new technique of paleomagnetism took advantage of the ability of rocks to record the geomagnetic field at the time of their formation, as we shall see in much more detail later. When I was a student at Cambridge in the 1950s, I attended lectures by Keith Runcorn, who was advocating the exciting possibility that Wegener's continental drift could be tested with paleomagnetism. At that time, it was generally accepted that given a roomful of geologists about 50 % would be strong advocates of drift and the other 50 % equally strong opponents of the idea. Runcorn's research group, first at Cambridge and subsequently at Newcastle, included Ted Irving and Ken Creer, whose contributions were to be pivotal. A second group, first at Manchester and subsequently at Imperial College led by Patrick Blackett and John Clegg also set out to use the geologic fossil compasses to navigate the continents back in time and to test the idea of continental drift. I was more familiar with this group because I had the good fortune to be John Clegg's nephew. He had taught me about the joys of working in science and largely decided the course of my life. Both groups found that many rocks recorded a magnetic field inconsistent with their present location and orientation: they must have moved over the face of the planet. For a while it seemed that the data might be explained by polar wander, whereby the earth as a whole moves in relation to the rotation axis. Critical data, particularly from Irving in the Newcastle group and from the London group eventually made it clear that the answer was continental drift.[6]

---

[6] Frankel (2012). This exhaustive study written with benefit of help from the central characters is an excellent presentation of the story.

In those days I had an opportunity to see how strongly scientists hold their views and how emotionally tied to them they can be. It seemed sometimes that adherents of different camps were near to coming to blows. However, out of this was to come the revolution in the earth sciences. Perhaps such strong interactions between scientists are an accompaniment of times of paradigm shift, as we would now say. Indeed, it has often been noted that opponents of the new ideas never change, they simply die unconvinced. For the most part, interactions between geophysicists are much more congenial now and so we may be in a period of consolidation and less excitement. Yet to some extent, in the magnetism of the moon, meteorites and asteroids, and in the events in the early solar system I sense some of the same passions as newer ideas upset the old order.

2.3 Summary

In this chapter, we have traced the growing understanding of the impact cratering on the earth and moon and taken a brief glimpse at the deep structure of the earth, age determination and paleomagnetism in the 1960s. They will all be central to our story.

References

Baldwin R (1949) The face of the Moon. University of Chicago Press
Boon JD, Albritton CC (1936) Meteorite craters and their possible relationship to cryptovolcanic structures. Field Lab 5
Bucher WH (1933) The deformation of the Earth's crust, Princeton University Press, Princeton
Chao ECT, Shoemaker EM, Madsen BM (1960) First natural occurrence of coesite. Science 132(3421):220–222
Dietz RS (1946) Geological structures possibly related to lunar craters. Popular Astron 54:465
Frankel HR (2012) The continental drift controversy: paleomagnetism and confirmation of drift. Cambridge University Press
Holmes A (1944) Principles of physical geology. Thomas Nelson and Sons, Edinburgh
Kopal Z (1962) Physics and astronomy of the Moon. Academic Press
Rutherford E (1904) Speech at the royal institution. Quoted in E.N. da C. Andrade, 1964, Rutherford and the nature of the atom. Anchor Books, Doubleday, p 80

Chapter 3
The Birth of the Space Age and Unmanned Missions to the Moon

After Sputnik and prior to manned flight, the Luna, and Explorer programs had taught us about the magnetic environment of the Earth with its Van Allen Radiation Belts, and shown that the moon unlike the earth had no detectable magnetic field. The Surveyor missions demonstrated that the Moon was an evolved body like Earth rather than a primitive body, so it looked as though there could be a core, in which a magnetic field might at one time have originated.

3.1 First Steps into Space and Early Heroes

Sputnik orbited the earth on Friday 4 October 1957 and innumerable accounts tell of the shock that it brought to the U.S. As one account goes, a group of American scientists learnt of Sputnik at reception at the Soviet Union's Embassy in Washington. The reception was in connection with a conference on rocket and satellite research. Apparently excitement had grown throughout the meeting as hints came for the Russians of a possible launch. At the reception Lloyd Berkner was informed of the successful mission and after clapping his hands for attention said,

> I wish to make an announcement. I've just been informed by the *New York Times*, that a Russian satellite is in orbit at an elevation of 900 km. I wish to congratulate our Soviet colleagues on their achievement.

The initial announcement of Sputnik was curiously muted in the Soviet press, but the next day, banner headlines extolled the virtues of a society that could achieve such a masterstroke.

The Russian program had been under the direction of Sergei Pavlovich Korolev (Fig. 3.1).[1] It is natural to compare Korolev, with von Braun and Goddard, although others might see the original trio of giants as Tsiolkovsky, Goddard and Oberth. As we shall see, this latter trio all provided remarkably prescient published work. However, it was Korolev, Goddard and von Braun, who were all three at similar stages in the development of liquid fuel rockets in the late 1930s.

---
[1] Korolev and Sputnik 1 Image courtesy of Rocket City Space Pioneers.

M. Fuller, *Our Beautiful Moon and its Mysterious Magnetism*,
SpringerBriefs in Earth Sciences, DOI: 10.1007/978-3-319-00278-1_3,
© The Author(s) 2014

Fig. 3.1 Sputnik1 and Sergei Pavlovich Korolev the "chief designer"

Fortunately, there are now excellent books that cover their stories and much of the following discussion comes from them.²

Born in 1882 Goddard was the oldest of the trio. During the First World War he developed a tube launched rocket for infantry use and was involved in the possible production of solid fuel rockets. By the 1920s he had already switched to liquid fueled rockets and claimed to have been the first to suggest rockets for high altitude flight and the multistage rocket. He had communicated with Oberth, who was to write a book "Rockets into Planetary Space" in 1923, which was favorably reviewed in *Nature*. The book acknowledged Goddard's work, but apparently Goddard felt it did not give him sufficient credit and relations between the two seemed to have been frosty from then on. Goddard became a celebrity in the 1930s, but then seemed to drop somewhat from sight until his death, when his wife Esther succeeded, through her lectures, to get his efforts recognized again. He had numerous patents that established key precedents. He had also recognized that if rocketry and space science were to be taken seriously, it would have to be backed up by rigorous mathematical and experimental work. Although he did not know at the time, he eventually learnt that Tsiolkovsky had provided much of the necessary analysis. In a remarkable publication of 1903, during the age of Zeppelins, Tsiolkovsky advocated the main developments to come in rocketry, backed up with numbers to prove his ideas on

---

² Clary (2003), Neufeld (2007), Harford (1997). All three books are gems among the many that trace detailed histories of these giants of the early space age and all have plentiful references to important earlier works.

calculating rocket thrust, demonstrating that rockets could operate in a vacuum, the velocity necessary to escape the earth, and multistage liquid fuel rockets.

In turning to von Braun, I must confess somewhat mixed personal feelings. As a 10 year old schoolboy, I had been on the receiving end of his V2 rockets in southern England at the end of WWII. Some have suggested that von Braun made a Faustian deal on his way to realizing his dream of space flight becoming a Nazi, and never showed much remorse over what had happened to the workers in the factories at Peenemunde and the underground plant in Kohstein Mountain. However, whether he really knew about the conditions there is far from clear. In England in those days, there was one good thing about V2's; if you were killed by one, you probably did not know anything about it, you certainly did not hear it coming, as one did with conventional bombs, or V1 s. When the first V2 arrived on September 8th 1944, there was no immediate recognition of what had happened, there was no air raid, but there was a 30' wide, 10' deep crater, houses destroyed, one person on the street disappeared and there were other casualties. Before long it became apparent that Goddard's ideas were being followed. Of course, the V2 s became fine fodder for the vilification of the Nazis, which was well deserved, but it is worth remembering that the total explosive power of the 3,000 or so V2 s launched was less than a single RAF night raid on a German city like Hamburg. On the positive side the V2 rocket technology was the bedrock upon which all subsequent rockets were based. Von Braun proved a visionary and an excellent manager of rocket teams.

Korolev had an amazing history, having been sent to the Gulag by Stalin in 1938 for "subversion in a new field of technology". This was when soviet military leaders, scientists, engineers, and ordinary citizens were sent by Stalin to the likes of Kolyma, where several thousand prisoners died each month. One of the most damning sins was for the paranoid Stalin to see you as a potential rival. Almost certainly this behavior of Stalin weakened the Soviet military and contributed to their initial failures against the German invasion. Somehow Korolev survived for 5 months and was recalled to Moscow because his case was to be reinvestigated. However, no travel arrangements were made and there followed an epic journey, which very nearly killed him, as he related one evening to his friends, the astronauts Yuri Gagarin and Alexei Leonov. After this, he was to be sent back to Kolyma with a reduced sentence from 10 to 8 years—a similar death sentence! Fortunately, due to intercessions of several prominent scientists he was sent to a sharaga. Here scientists experienced less harsh conditions, so that they could continue their work. It is a constant source of amazement that prisoners remained loyal to the Soviet Union and continued their work. As Sakharov noted in his memoirs (Sakharov 1990), an indication of the insanity of the situation was the story of one professor, who was protesting his innocence to Beria, whereupon Beria replied,

> My dear man. I know that you are not guilty of anything. Get the plane in the air and you will go free.

That would I suppose provide considerable motivation in the sharagas, but even more remarkable is that after release back into society, men like Korolev remained devoted to their country and worked feverishly for it. While he obviously had important colleagues, he managed the programs that put Sputnik in orbit, followed by Sputnik 2 with the dog Laike, the first astronaut Gagarin, the first two man vehicle, the first three man vehicle, the first woman in space and the first space walker in Alexei Leonov. During his life Korolev was primarily acknowledged as the Chief Designer, possibly for security reasons. However, with his death his achievements were finally recognized and he was given a state funeral, in the Hall of Columns of the House of Unions, with solemn music by Tchaikovsky, Beethoven and Chopin. The ribbons on the casket proclaimed at last "to the outstanding Soviet scientist, twice Hero of Socialist Labor, Lenin Prize winner, the Academician Sergei Pavlovich Korolev"

With the launch of Sputnik, the Jodrell bank radioastronomy facility of Manchester University, in the quiet Cheshire countryside, now played an important role by tracking Soviet space vehicles. I had known Jodrell bank as a schoolboy, in the late 1940s and remember very clearly old military searchlights with Yagi arrays mounted on them. At this point, my uncle Johnnie Clegg was doing his thesis research in radioastronomy with Bernard Lovell. Lovell had had a distinguished career working on a number of military applications of radar, at an age when modern scientists would probably be looking for a post doc. He played a key role in airborne aircraft interception equipment, used in night fighters and in H_2S, the curiously named, airborne radar system that revolutionized night bombing and blind bombing through cloud by providing radar maps of the target area as they flew over. Whereas he was happy to have helped defend his country, he expressed concern over the offensive role of the blind bombing. With the end of the war Lovell went to Manchester University to work with Blackett in radioastronomy. The Yagi arrays on old military searchlights were to metamorphose into a stationary dish antenna with which important results were obtained, but the construction of the giant steerable dish Mark 1 radio telescope (Fig. 3.2) was a far larger endeavour and suffered from cost overruns and difficulties in staying financially solvent. However, when it was found that it could track space vehicles its problems were soon to be over. The telescope was used to track the American and Russian moon rockets and Pioneer, which helped to establish the radiation belts of Earth (see below). With Pioneer 5 it actually played an active role in relaying commands to the vehicle instead of simply passively tracking them.

Jodrell Bank was seen as a British contribution to the new space age and Lovell had a call from Lord Nuffield one day, who asked,

> "How much do you still owe on that telescope of yours? I want to pay it off" Lovell recalled that he tried to start thanking him, knowing that the sum was somewhere in the region of £150,000, but he did not get a chance. Lord Nuffield went on,

> That's alright, my boy, you haven't done too badly.

This was the end of financial problems that had dogged the instrument from the earliest days. It became the Nuffield Radio Astronomy Laboratories and it was free to play its true role in radioastronomy.

3.1 First Steps into Space and Early Heroes

Fig. 3.2 Mark 1 antenna of Jodrell Bank (Courtesy Peel 1991)

To return to the Russian program, within a month a second satellite with a mass five times greater than Sputnik 1 and carrying the dog Laika was launched and was tracked by the Jodrell Bank group. It appears that Sputnik 2 was something of a rushed job in response to Kruschev's request for a speedy second space spectacular. Sputnik 2 stayed in orbit for 200 days. Laika, a stray, that achieved such immortal fame perished on about day 4, when the heater control system failed. Sputnik 2 broke up on reentry and was never intended to be recovered, so Laika's fate was doomed in any case. For dog lovers, like my wife, we note that one of the Russian scientists publicly expressed his regrets that Laika had been killed and said that the mission should never have been carried out.

The U.S. was developing the capability to launch a satellite into orbit with the Explorer program and might have launched before the Soviets in different circumstances. A positive side of the response to Sputnik was a dose of reality in the form of a national recognition that good though U.S. science and technology was, other modern advanced countries were also capable of technological triumphs. Another benefit was the Eisenhower administration's major effort to strengthen science and engineering education and research, from which I, and a generation of scientists in the U.S. gained support for our research.

A major achievement of the US Explorer program was the discovery of what are now known as the Van Allen Belts, named after their discoverer—James Van Allen. A Geiger counter and an altimeter had been placed on Explorer 1 and revealed the radiation belts. The magnetosphere was increasingly well defined with the bow shock of the magnetosphere recognized as the outer margin of the magnetosheath within which the solar wind particles are stood off by the earth's magnetic field. It had long been known the streams of particles flowed from sun to earth and caused magnetic storms on earth by their interaction with the

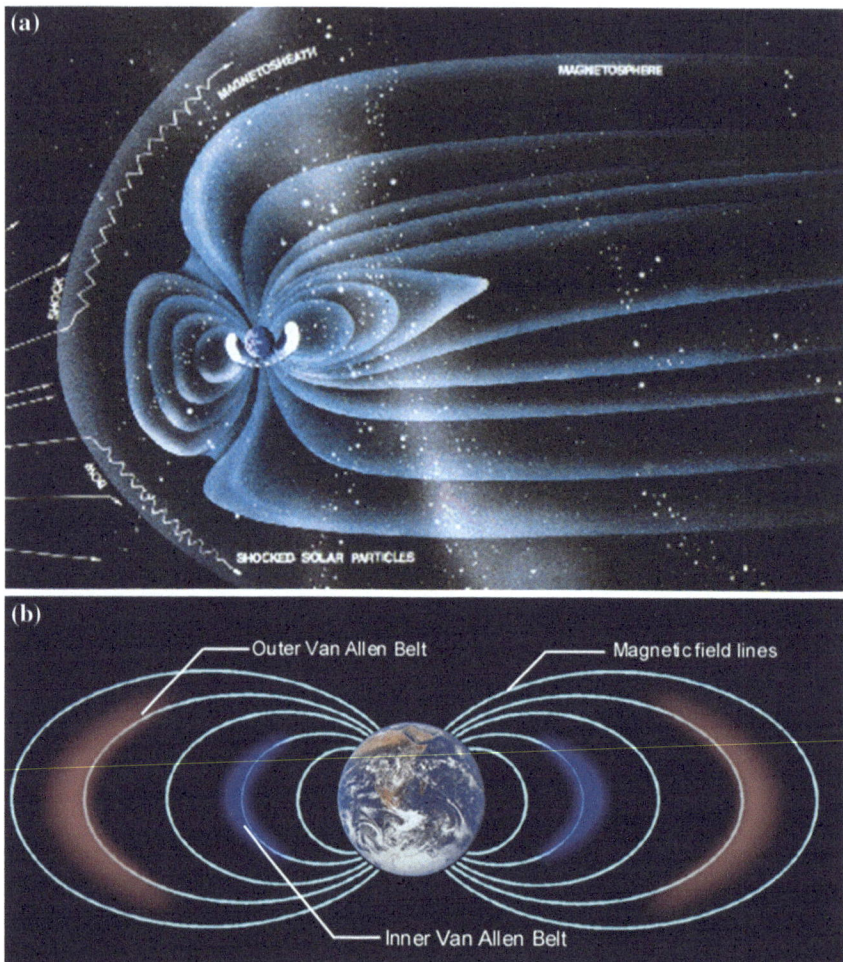

Fig. 3.3 a The magnetosphere and **b** Expansion of central region to show the Van Allen radiation belts

magnetosphere.[3] The earth is protected from the solar wind particles by the magnetosphere. The particles within the belts spiral around field lines in response to the Lorentz force they experience (Fig. 3.3).[4]

---

[3] The occurrence of a solar flare in 1859 followed a day later by a magnetic storm suggested to each of Carrington and Hodgson independently that a stream of particles was flowing from the sun to the earth and that the strength was dependent upon events on the sun. Half a century later in 1910, Eddington made the same suggestion in connection with a study of Comet Morehouse. The term solar wind came from Parker, who had developed a model for the escape of supersonic particles from the outer corona of the sun.

[4] NASA on line images.

3.1 First Steps into Space and Early Heroes

The Lorentz force arises when a charged particle is in motion in a magnetic field. The strength of the force is in fact used to define the intensity of a magnetic field. There is a simple way to visualize the effect of the force. This follows from some mathematics that we will skip. If you hold your right hand with the thumb vertical to represent the direction of motion of a positive charge, and the fingers extended to represent the direction of the magnetic field, then the Lorentz force will be horizontally out of the palm. If the charge is negative the force will be in the opposite direction, into the palm. The Lorentz force is perpendicular to both the velocity and the magnetic field directions. The result of this force is that the particle will spiral along a field line, which is exactly what happens to the particles in the radiation belts. There is one additional effect—if the field strength increases, then the pitch of the spiral will steepen and eventually the particle reverses its motion, moving away from the strong magnetic field source. This happens as the field lines approach the stronger fields near to the poles. The particles are then reflected to spiral backwards and eventually forwards again along the geomagnetic field lines and so giving the geometry of the radiation belts. The outermost belt is populated with electrons and the inner with a mixture of electrons and protons. The continued documentation of the radiation belts and of the magnetosphere was achieved with a combination of results from Explorer, Pioneer, and Luna unmanned missions.

3.2 Unmanned Missions to the Moon and the Beginnings of Apollo Program

The Apollo program started in July 1960 in response to President Kennedy's challenge to put a man on the moon and to return him safely within the decade. Looking at the record of U.S and Soviet Space shots in the late 1950s and early 1960s, it is clear how difficult a task NASA had been set. In 1958, the first 3 Soviet unmanned spacecraft with missions to impact the moon failed with booster rocket explosions within 250 s of launch. In the next year there was one more failure at the third stage. Luna 1 escaped earth's gravity, but missed the moon, Luna 2 provided the world with the first lunar impact, and Luna 3 gave us the first picture of the far side of the moon. Even following these success, continued Soviet failures went on into the mid-1960s. The U.S. had similar problems in the same period with the Pioneer spacecraft, were launched on modified ICBMs with the aim of lunar impacts. The first, which was managed by the United States Air Force blew up on the pad. The second and third burned up on reentry far short of the moon, but with Ranger 7, 8, and 9 successes were achieved. Evidently getting a man to the moon and returning him to earth was not going to be easy, but major progress was being made.

By the second half of the decade of 1960s both the U.S. and the Soviets had successfully landed vehicles on the moon. Surveyor 7 included establishing

Fig. 3.4 a Pete Conrad and Surveyor 3 on Apollo 12 **b** Surveyor landing sites

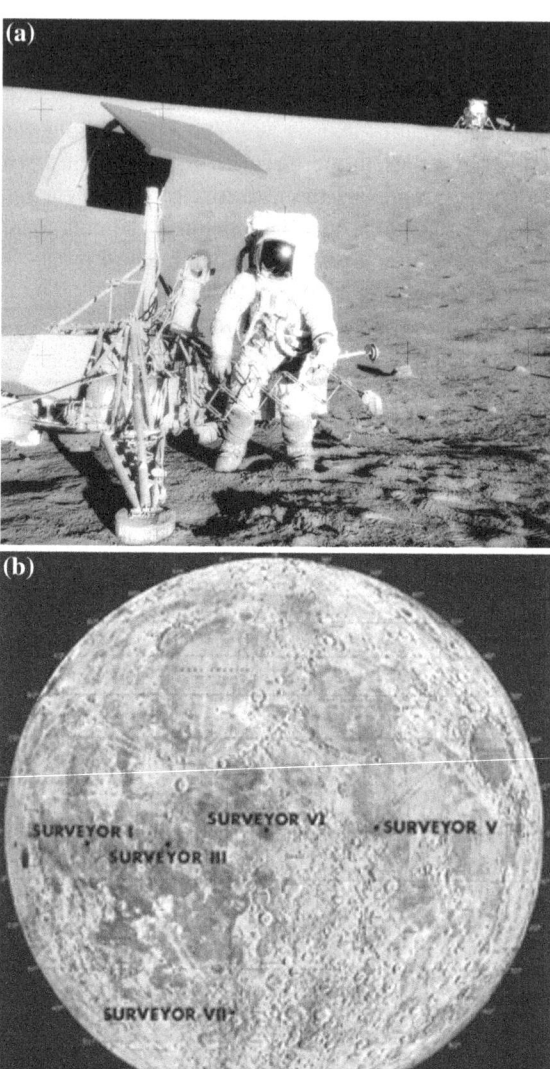

surface mechanical properties, temperatures, and electromagnetic properties. The first landers also conclusively established that space vehicles would not disappear in the lunar dust, as had been suggested by some. Surveyor carried equipment to determine the composition of the soil at the landing sites and successfully measured this at 5, 6, and 7 sites (Fig. 3.4).[5]

---

[5] Images courtesy of NASA.

3.2 Unmanned Missions to the Moon and the Beginnings of Apollo Program

This equipment was an Alpha Particle Scattering spectrometer based on one of the most famous experiments in all physics. In 1911 at Rutherford's suggestion, Geiger and Marsden had bombarded a gold foil with α-particles—the two protons and two neutrons of the Helium nucleus. Most of the particles passed straight through the foil, but to general surprise, a few were scattered at strong angles from the foil. Hence a few α-particles had experienced very strong repulsive forces, while most sensed nothing and passed straight through the foil. This argued for a dense nuclear source and an atom that was mostly empty space, giving rise to the modern view of the atom.

To determine chemical composition using this phenomenon, one takes advantage of the fact that when an α-particle strikes a nucleus of an atom the impact causes the nucleus to recoil. This recoil takes energy from the α-particle. The amount of energy taken in recoil depends upon the mass of the target atom and so if the energy of the returning α-particle is measured, the mass of the nucleus it struck can be obtained. The three Surveyor sites that were analyzed were Mare Tranquillitatis, the small Mare Sinus Medii and Tycho in the southern highlands. Oxygen $\sim 60\,\%$ and Silicon $\sim 17\,\%$ dominated at all sites, with no Carbon. The Iron and Titanium at the Mare sites were higher than at the highland site, consistent with the Mare being basaltic lava flows.

The story of unmanned missions to the moon was far from over at this point. Indeed, when Apollo 11 was on the way to the moon, a Russian mission to return sample was also in transit. Unfortunately, it crashed into the moon, but shortly after sample was returned for the first time by an unmanned lander—Luna-12. A series of similar missions helped extend the lunar sample collection from the Apollo missions to different regions on the moon. Important orbiters since Apollo, such as Clementine, Lunar Prospector, Lunar Orbiter, Lunar Reconnaissance Orbiter (LRO) and Grail have contributed to our understanding of the moon and we will use their results when we discuss post-Apollo science. Japan's Selene (Kagayan) orbiter has already returned some of the most detailed images of the lunar surface. Soon China and India will join the nations with lunar landers and there may again be lunar exploration by manned or unmanned missions to the surface.

3.3 Summary

With the unmanned missions and particularly with the Surveyor results, it became clear that the moon was an evolved body with lavas on it surface not too different from those here in the Hawaiian chain. Whether there was a core was not yet clear.

References

Clary D (2003) Rocket man—Robert H. Goddard and the birth of the space age, Hyperion Books
Harford J (1997) Korolev—How one man masterminded the Soviet effort to beat America to the moon
Neufeld MJ (2007) Von Braun, dreamer of space engineer of war
Sakharov A (1990) Andrei Sakharov Memoirs, Alfred A. Knopf, New York

Further Reading

Courtesy Peel M (1991) Jodrell bank centre for astrophysics, University of Manchester, See also Lovell B., "Astronomer by chance", McMillan Books, London

Chapter 4
Apollo: Getting to the Moon

The Apollo program was preceded by Mercury (1961–1963), Gemini (1964–1965), and by the Russian, Vostok (1961–1963) and Voshkod (1964–1965) Soyuz (1967–) programs. We develop the history of the events in the Apollo program leading to Apollo 11, with a brief summary of the subsequent Apollo missions. One of the most famous photos of the Apollo missions was taken before any lunar landings. It came from the Apollo 8 crew as they orbited the moon (Fig. 4.1).[1] For many, it taught us what a small and precious pearl the earth is. For some, this was the greatest lesson of Apollo.

4.1 Apollo Background

To understand Apollo, we need to go back to the Mercury and Gemini programs. The Mercury project began in 1958 only just over 1 year after Sputnik. The challenge was to build a capsule that would protect the occupant from the vacuum of space and the temperature extremes, the cold during orbit and the heat during reentry.

The capsules were tested with monkeys and chimpanzee of which the most famous was Ham, a chimpanzee. The first Mercury vehicles were launched on Redstone rockets from Werner von Braun and the group at Huntsville (Fig. 4.2).[2]

The seven astronauts of the early missions had been selected from a group of about 100 military pilots and the number 7 was included on each of the Mercury spacecraft as a token of the cooperative spirit of the group. All of the seven flew in Apollo, Gemini, or Mercury programs, with the exception of Deke Slayton, who was diagnosed with a heart condition, although he too flew later in the Shuttle program. Alan Shepard was the first American in space. He was a naval aviator and had been to the Navy test pilot school at Pautaxent River, Maryland, where he

---

[1] NASA Apollo 8, Gallery.
[2] Photos courtesy of NASA.

Fig. 4.1 The earth from the moon as photographed by Apollo 8

Fig. 4.2 Redstone and Atlas launches

had been involved in tests of important advances for Naval Aviation. He reported that shortly before launch as he sat atop the rocket he said to himself

> Don't fuck up Shepard…

4.1 Apollo Background

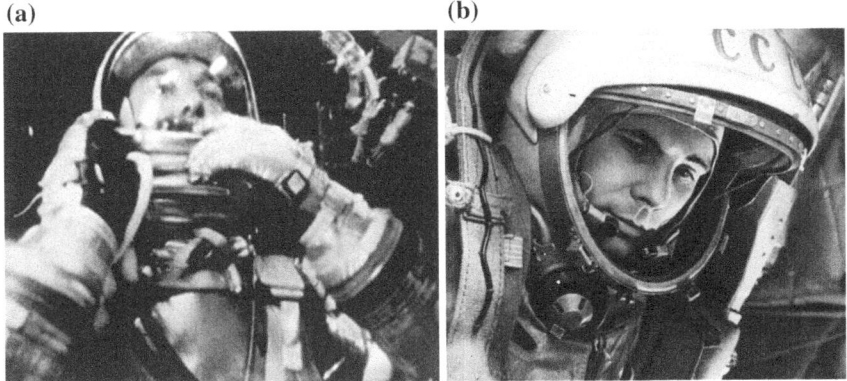

Fig. 4.3 **a** Alan Shepard in Freedom 7 and **b** Yuri Gagarin in Vostok 1

A misquoted version of this has become known as Shepard's prayer and is dear to every pilot no matter how experienced or how amateur. He was a cool customer and had a good sense of humor. On being asked what he was thinking about while waiting for lift off he replied,

> The fact that every part of this ship was built by the low bidder.

His mission was initially planned for October 1960, but was postponed several times until May 5th 1961. By this time, Yuri Gagarin had orbited the earth (Fig. 4.3).[3]

In his official statement Gagarin reported that,

> On the 12th of April, 1961, the Soviet spaceship-sputnik was put in orbit around the Earth with me on board …there was a good view of the Earth which had a very distinct and pretty blue halo. It had a smooth transition from pale blue, blue, dark blue, violet and absolutely black. It was a magnificent picture.

Sadly Gagarin was killed in the crash of MiG 15 only a few years later on 27th March 1968.

Shepard's sub-orbital flight was followed by Grissom's and then came John Glen's orbital flight in Friendship 7. By that time I had arrived in the U.S. and like everyone else was enthralled by the space program. No matter what one's religious persuasion, or lack of it, all shared with

> Godspeed, John Glen!

The mission had a terrifying conclusion with fears that there might be trouble with reentry because the heat shield had been compromised. The tension was made all the more unbearable because during the reentry there are 4 min of broken communication when the plasma generated during reentry cuts off radio contact, but all was well and the country heaved a concerted sigh of relief.

---

[3] See footnote 2.

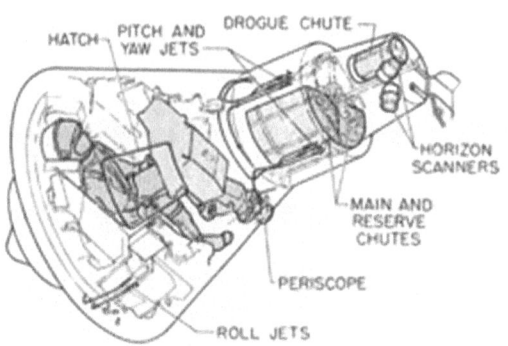 

Fig. 4.4 Capitals for Mercury and Friendship

The remaining 3 Mercury orbital missions were flown by Carpenter in Aurora 7, Schirra in Sigma 7, and Cooper in Faith 7. The program had succeeded in its aims of testing the response of the astronauts to conditions within the capsule during orbital missions. Recovery was by splashdown, which could be life threatening if there were a malfunction of the capsule, or if the landing was far from the rescue vessel. In the Mercury program there were some major misses, but the last two flights landed within 10 km of the rescue vessel (Fig. 4.4).

The second U.S. manned space program was Gemini during which two man vehicles were used for docking training, and longer stays in orbit. The first manned flight, Gemini III, was by Grissom and Young. Gemini IV included the first space walk with White outside of the capsule for 22 min. The remaining Gemini flights continued to develop skills for docking. Gemini VIII was notable for the successful handling by Armstrong and Scott of uncontrollable spinning of an unmanned Aegena stage with which they were docked. The Gemini was undocked and an emergency landing achieved.[4]

The Apollo program ran from 1961–1972. It was conceived in the last days of the Eisenhower administration. In November, John F Kennedy was elected after his campaign in which much was made of missile gaps and the need for the U.S. to surpass the Soviet Union. After Yuri Gagarin's flight, Kennedy requested that his Vice President Johnson look into the space program and see what were the chances of catching up with the Russians. A somewhat negative report came from Johnson concerning the level of effort being made, but it included the opinion that there was a possibility of getting to the moon first, since the event was a long way off. Kennedy decided on a bold move and on May 25th made his famous speech:

---

[4] Diagram and Photo courtesy of NASA.

4.1 Apollo Background

> I believe that this nation should commit itself to achieving the goal, before this decade is out, of landing a man on the Moon and returning him safely to the Earth. No single space project in this period will be more impressive to mankind, or more important in the long-range exploration of space; and none will be so difficult or expensive to accomplish.

The financial outlay and a frequently quoted number would be enormous, around $\sim$ \$20 billion in 1967 dollars. To a generation growing used to discussions of trillions of \$'s, financial companies worth hundreds of billions and gambling with billions this may not seem so much, but in the 1960s.

> A billion here a billion there and pretty soon you were talking about real money

In the supposed immortal words of a famous speaker of the House.[5]

The mission was clearly immensely difficult, with no agreement at the outset how it should be done. Remember that these plans were being made after only one sub-orbital mission. The most straightforward approach was to fly a rocket to the moon and to cut down weight requirements on return by leaving its landing stage on the moon. However, this would still need a new generation of much more powerful rockets and perhaps even multiple launches. There were also ideas for assembly in earth orbit, or a rendez-vous on the lunar surface. One an initially unpopular view was eventually proved right. This was for a lunar orbit separation of the lander from the command module, followed by rendez-vous with the lander in orbit after completion of the lunar surface work. Such an approach had been advocated by Tsiolkovsky at the beginning of the century! Even the most superficial look at the story of Tsiolkovsky makes clear one is in the presence of genius. He seems to have recognized most of the potential problems likely to arise in space flight with the solutions to them and predicted among other things, liquid fuel multistage rocket boosters, tethered space walk suits, near earth space stations, and rocket guidance by firing nozzles, to say nothing of his mathematical analysis of reaching space by rocket.

To return to the approach that was eventually used, its key point was that reduction in weight requirements meant that the mission could be launched on a single rocket. Eventually after about a year the Langley Group was coming around to support the idea and finally the Marshall Space Center acquiesced.

4.2 Apollo 1 to 11

At the outset, the Apollo program suffered one of the worst of all NASA accidents. On 27 January 1967, Virgil "Gus" Grissom, Ed White and Roger Chaffee were killed in a terrible fire on the Apollo 1 launch pad, during a launch simulation. The immediate problem was that the gas in the capsule was pure oxygen and a spark had

---

[5] The Dirksen Congressional Center. Whether Speaker Dirksen ever actually said this is not clear, but it is generally ascribed to him.

Fig. 4.5 Apollo 1 crew. *Left* to *right* White, Grissom and Chaffee

occurred. The loss of all three was a national tragedy and Grissom with his experience, knowledge from working with the engineers, and his reputation for speaking his mind must have been a particularly devastating loss to the program (Fig. 4.5).[6]

A similar tragic accident in the Russian program had killed Cosmonaut Valentin Bondarenko. The cosmonaut near death acknowledged that it was his fault because a pad he had discarded landed on a heater and started the fire.

> It's my fault, I'm so sorry… no one else is to blame.

These were the near final words of this Russian hero.[7]

The Apollo 1 accident resulted in a major overhaul of the capsule design, with astronaut Frank Borman leading the interaction with the engineers. The modifications delayed the first Apollo Mission for over a year, but permitted safe operation throughout the remaining Apollo missions. With the new start a series of missions rapidly followed each other using a variety of Saturn Launch Vehicles of which a Saturn V is illustrated (Fig. 4.6).[8] The V refers to the five F-1 engines that power the first stage. The first stage (S-IC gets the rocket to an altitude of ∼38 miles, after burning ∼2.5 min, it then separates and burns up in the atmosphere. The second stage (S-II) has 5 J-2 engines that burn for ∼6 min taking the assembly to ∼115 miles, where it too is jettisoned. The third stage (SIVB) with 1 J-2 engine burns for ∼2.5 min and achieves orbital velocity ∼17,500 mph. It is shutdown with fuel remaining, and is later used for trans-lunar injection. The record of these Saturn launch vehicles is remarkable with no failures in any of the launches.

The Apollo 7 crew of Schirra, Eisele and Cunningham launched on October 11th 1968 and performed a number of firsts including docking with the SIVB stage, from which they had separated shortly after entering orbit. They successfully completed piloting and navigation tasks. They also provided the first live TV from space. The crew enjoyed the space available in the vehicle and reported no untoward effects on them or the capsule. The Apollo 7 splashed down less than 2 km from the planned

---

[6] Photo courtesy of NASA on line image ESA/alldayru.com.
[7] Korolev, James Harford (1997)
[8] NASA Apollo 9 Gallery.

4.2 Apollo 1 to 11

Fig. 4.6 Apollo 6 the final test of the Saturn V rocket. Command module (*CM*), service module (*SM*) and lunar module (*LM*)

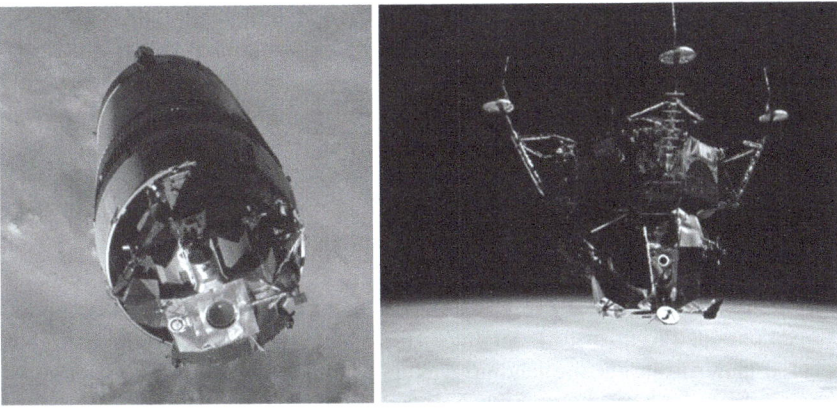

Fig. 4.7 The Apollo 9 lunar module awaits docking and extraction from the SIVb stage and in earth orbit after extraction

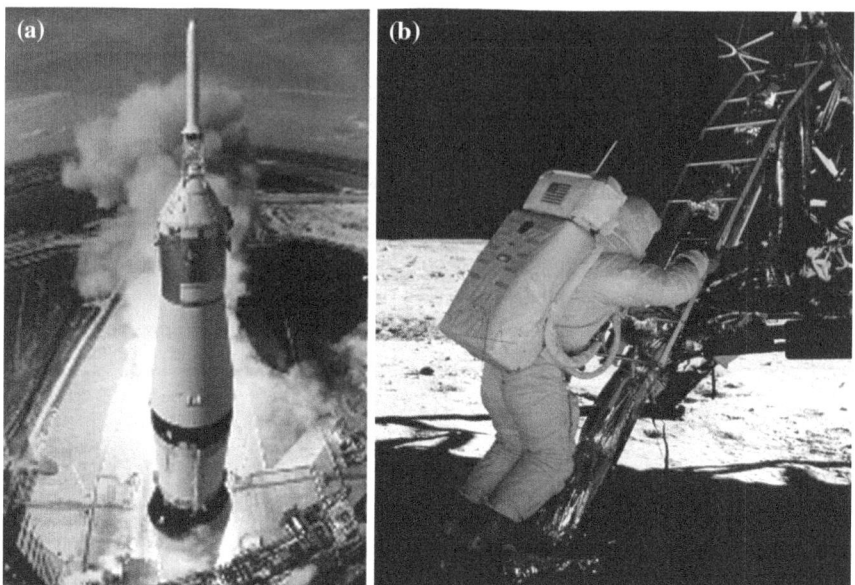

Fig. 4.8 a Apollo 11 lift off from launch tower window showing escape assembly and the command module on *top* and **b** Buzz Aldrin on the LM footpad

point in completion of a highly successful mission, despite all three crew members having head colds during their time in orbit. There was some friction between the crew and ground control and the crew never flew again. They had to wait until 2008 before being accorded the recognition that they so richly deserved.

With Apollo 8 (Borman, Lovell, Anders) orbiting the moon was achieved. Then with Apollo 9 (McDivett, Scott, Schweikert) extracting and docking with the Lunar Module was tested in earth orbit (Fig. 4.7), and finally in lunar orbit with Apollo10 (Stafford, Young, Cernan). The Apollo 10 Lunar Module with Commander Stafford and Lunar Module Pilot Cernan came to within 50,000 ft of the lunar surface before climbing back up to dock with the Command Module (Young). With these manoeuvers completed successfully, all was ready for the landing mission. Within one year, Apollo 7, 8, 9 and 10 had been completed with the launching and safe recovery of 12 astronauts—a truly remarkable achievement. Now it was time for Apollo 11 and the meeting of President Kennedy's challenge.

Apollo 11 launched at 9.30 am, on July 16th 1969, carrying Commander Neil Armstrong (C), lunar module Pilot Buzz Aldrin (LMP) and Command Module Pilot Michael Collins (CMP) (Fig. 4.8a). The flight plan followed that of Apollo 10, and continued to its historic landing. The launch and 3-day travel to the moon appeared uneventful, but the landing was quite different. Early in the landing, a program alarm went off, but quickly Steve Bales the flight controller responsible for the LM computer came back with,

Roger, we're Go on that alarm

4.2 Apollo 1 to 11

This let the crew know it was OK to keep going. The problem had been a computer overload and postponement of some non-critical tasks. A little later at three thousand feet above the surface, another similar alarm went off and again the same response came from Houston. At that point, Armstrong looked at the projected landing site and found that they were heading for a very large crater with plenty of boulders (West Crater). To avoid this they manually extended downrange and at touchdown had less than 30 s of fuel left.

Houston, Tranquility Base here, The Eagle has landed!

This was followed a little later by

Tranquility: That may have seemed like a very long final phase. The auto targeting was taking us right into a football-field-sized crater, with a large number of big boulders and rocks for about one or two crater-diameters around it, and it required flying manually over the rock field to find a reasonably good area.

This was not the first time that Nell Armstrong had demonstrated what a remarkable pilot and man he was. As we saw above, he and David Scott had managed to overcome a thruster problem on Gemini 8, successfully undocked and returned safely to earth. Armstrong had also successfully ejected from the flying bedstead (Lunar Landing Research Vehicle LLRV) with seconds to spare and quietly returned to his desk to continue his paperwork! On the moon he now uttered his immortal words,

That's one small step for (a) man, one giant leap for mankind

The first task on the lunar surface (Fig. 4.8b)[9] was to collect the contingency sample of scoops of soil and rock, which turned out to weigh about 1 kg. It was collected in view of cameras close to the LM. This was followed by a traverse to the south and east of the LM involving a total distance of about 60 m to Little West crater. The collection of the bulk sample, which consisted of 15 kg of rock and soil took place during this traverse of about 400 m to West Crater. The surface material produced by the aeons of bombardment was named the regolith and was 3–6 m deep here. In addition to collecting samples the first of the Apollo Scientific Experiment Packages was deployed including the Passive Seismic Experiment to detect moonquakes and the Laser Ranging Retroreflector, to provide accurate measurement of the earth moon distance.

4.3 Brief Summary of Remaining Apollo Missions

Now we will rapidly review the remaining missions from the point of view of sites (Fig. 4.9),[10] aims and equipment used, before going to the scientific results. Apollo 12, with astronauts Conrad (C), Gordon (CMP) and Bean (LMP) was launched

---

[9] NASA, Apollo 11 Gallery.
[10] Courtesy of NASA.

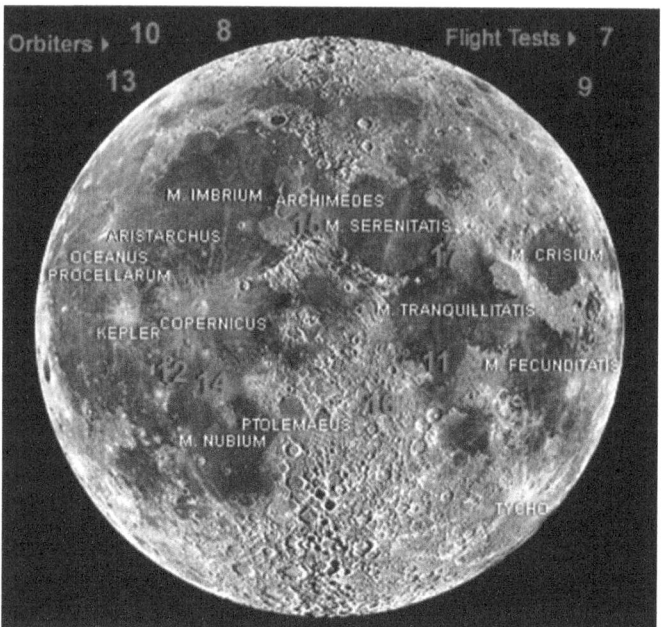

Fig. 4.9 Apollo missions and landing sites

14th November 1969 and executed a pinpoint landing in Oceanus Procellarum within 183 m of the Surveyor 3 craft that they were to visit. A seismometer and a magnetometer were among equipment deployed.

Apollo 13 was supposed to land in the Fra Mauro region, but an explosion of oxygen tank 2 forced a return to earth without a landing on the moon. In this the LM was used in a lifeboat mode to provide living space for the crew during the lunar earth transit. The story has often been told in print, documentary and movie form, but suffice it to say here that the safe return of astronauts Lovell (C), Swigert (CMP), and Haise (LMP) was a testament to U.S. ingenuity in general, and to the skills of NASA's astronauts and ground control in particular.

Apollo 14 with astronauts Shepard (C), Roosa (CMP) and Mitchell (LMP) landed in hilly terrain north of the Fra Mauro crater. The mission was designed to sample the ejecta from the Imbrium Basin. The inclusion of a two wheeled cart called in NASA's penchant for odd terminology the Modular Equipment Transporter, or MET, increased the range of exploration and permitted the use of a portable magnetometer.

The Apollo 15 crew of Scott (C), Warden (CMP) and Irwin (LMP) went to the Hadley Rille region to sample the Imbrium rim, mare flows from the Imbrium plains, and to investigate the structure of the rille. Extensive lava plains were exposed at the landing site and the massive Imbrium rim towered over the site

4.3 Brief Summary of Remaining Apollo Missions   43

4 km to south. The regolith is only 5 m thick throughout much of the site and absent near the edge of Hadley Rille, in which the layered basalt flow structure could be seen. This mission was the first to be equipped with the Lunar Rover Vehicle (LRV), which greatly expanded the range of exploration.

Apollo 16 had a curious history. It was sent to investigate the Cayley Plains and Descartes formations, which were thought before the mission to be volcanic plains, but were found to be highland material. The crew of Young (C), Mattingly (CMP) and Duke (LMP) deployed a portable magnetometer, a seismic experiment involving mortars and thumpers, a passive seismometer, and heat flow equipment.

Apollo 17 landed on the boundary between highlands and mare basalts in the Taurus-Littrow valley near the rim of the Serenitatis Basin.[11] With Commander Cernan and Command Module Pilot Evans, for the first time a professional geologist, Schmitt, was included as the Lunar Module pilot. Using the LRV traverses of kilometers were completed. Note the scale in the Apollo 11 photo (Fig. 4.10a), the tracks of the Armstrong and Aldrin can be dimly made out leading from the LM 10 s of m east to a small crater and in other directions to the ALSEP. The Apollo 11 photo (Fig. 4.10a) shows traverses of ~ 100 m, but with the Lunar Rover on Apollo 17 traverses of kms were made possible (Fig. 4.10b). The Apollo 11 photo also shows ejecta from West Crater that Armstrong had to overshoot to land safely.

4.4 The Training of Apollo Astronauts as Geoscientists

It is sometimes overlooked how important geological training was for the astronauts, so that they could take full advantage of their scientific opportunities. This is not in anyway to diminish the astronaut's role as pilots. Once when I was at the Cape, I saw some of the astronauts training manuals. That brought home to me just how much they had to learn. However, they were all very experienced pilots (top of the pyramid) and one suspects that to master this aspect of the missions, despite its awesome requirements was not overwhelming, although there would be plenty of worrying moments.

To become a field geologist for those who landed on the moon, or a photo-geologist in the case of the command module pilots was quite a different task. Despite this, they all did excellent jobs as geologists, although only Schmitt on Apollo 17 was trained as a geologist before the missions. The initial Apollo missions were obviously concerned primarily in the words of President Kennedy of landing a man on the Moon and returning him safely to the Earth. Increasingly, and reaching a peak in the Apollo 15, 16 and 17 missions, the contributions of the astronauts on the lunar surface as field geologists and the Command module pilots as a photogeologists became central to the success of Apollo. Their training in this

---

[11] NASA, Apollo 11 and Apollo 17 Galleries.

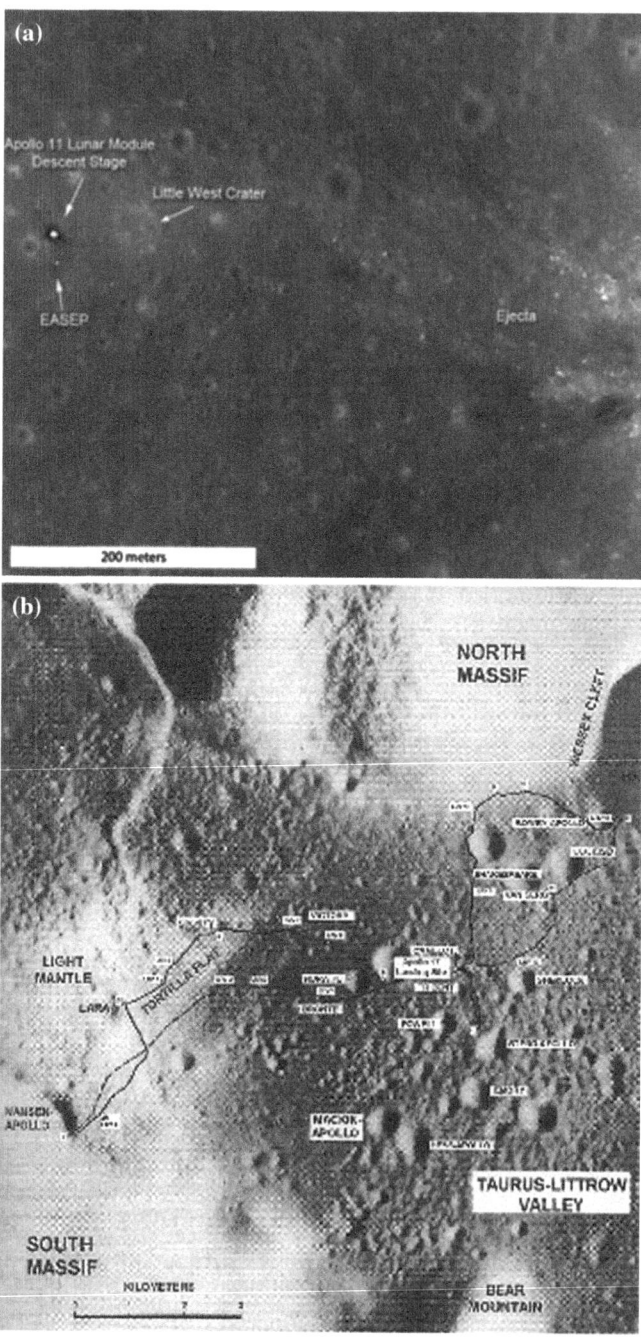

Fig. 4.10 (a) Apollo 11 site showing traverses of ~100 m and (b) Apollo 17 site showing traverses of km

area was in the hands of a number of geologists. Foremost among these in the early days was Gene Shoemaker, who we have already met in this story with his contributions to impact crater studies.

An early step was to take the Gemini astronauts on a field trip around Flagstaff including a trip to Meteor crater and then to classroom studies. At this point, or soon after, there appears to have been some antipathy between US Geological Survey and NASA as to the training of the astronauts, and indeed analysis of the samples. However, cooler heads prevailed and an excellent cooperative program developed which laid the foundation for the fieldwork to be carried out so well by the Apollo crews. This included extensive field trips and classroom time. Not only would the astronauts be collecting samples and making field observations, but they would be setting up sophisticated instrumentation on the lunar surface. They achieved all this with very few mishaps. Had it not been for their skilled handling of the magnetometers on the lunar surface the magnetic story would have been much weaker.

The Command Module pilots were to be the photogeologists, or perhaps we should say photoselenologists, during the Apollo mission. They had a difficult time because it was cramped in the module and particularly hard to get access to the window from which they would be photographing. The job was to scout possible forthcoming Apollo sites and to catch targets of opportunity that could cover specific requests from the various scientists.

4.5 Summary

This brief summary of the early manned missions culminating in Apollo covers a period of great worldwide excitement over the space program that took place at a time of social and political turbulence in the USA.

Chapter 5
Advances in Lunar Science with Apollo

In this chapter we review the results from the Apollo missions. With the availability of the samples, we immediately saw that they consisted of basalts, similar to those on earth, and breccias, which were the result of impacts. We also came face to face with the puzzle of lunar magnetism. Not only did the samples carry remanent magnetization (NRM), but surface magnetometers showed that there were fields at the various sites far larger than predicted by the earlier satellite measurements. Finally magnetometers on sub-satellites revealed that large regions of the lunar crust were magnetized. Yet there was no active lunar dynamo field at present. How did this magnetization arise?

5.1 Apollo Scientific Results

The Apollo 11 Lunar Science Conference was held in Houston from Jan. 5th to 8th 1970, roughly 6 months after the mission.[1] I attended with Professor Nagata and eagerly awaited the results to be presented. It was obvious that many of the key questions would at last be answered. Using the same method as Patterson had for his determination of the age of the earth, Tatsumoto and Roholt reported, an age for the moon, and with minor later modifications, we now had an age of ~4.53 Ga, which placed it some 10's of million years younger than the earth. We immediately saw the nature and composition of the lavas and the lunar surface, or regolith, which was the product of the aeons of surface bombardment. We also saw the magnetism of the returned samples. Let us now look in a little more depth at some of these results from Apollo 11 and those from later missions (Fig. 5.1).

The Apollo 11 samples consisted of basaltic lavas, breccias, and soil. The basalts were mostly fine grained with some medium grained. The ages of the basalts from different methods in different laboratories converged on around 3.7 Ga. The soil consisted of crystalline fragments of the basalts, glasses and some exotic material, which was recognized to be from the highlands. The breccias were

---

[1] *Apollo 11 Lunar Science Conference Issue*, Science, 1970.

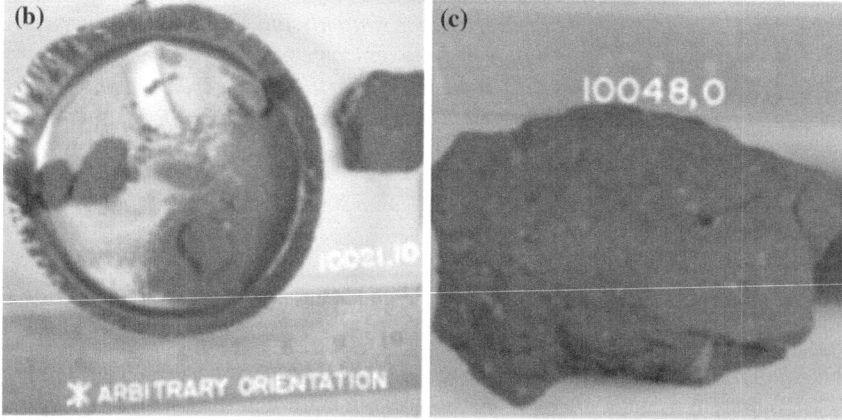

Fig. 5.1 Apollo 11 hand samples (scale ∼10 cm). **a** mare basalt 10020, **b** poorly consolidated regolith breccia 10021 and **c** well consolidated regolith breccia 10048

clearly shock welded soil, varying from some that barely had any cohesion (Fig. 5.1b) to well indurated rocks (Fig. 5.1c).[2] They were called regolith breccias. We had a sample from 10048 and knew that it was a solid rock, which appeared to carry some stable remanent magnetization.

Many of the rocks showed evidence of bombardment by hypervelocity particles yielding craters down to dimensions of a few mms (Fig. 5.2).[3] The impacts extended from the scale of microns to hundreds of kms. In addition, the lunar samples also carried a record of ionizing radiation arriving at the moon. The lunar surface was covered with a layer of material that had been generated by the

---

[2] NASA Sample Gallery.
[3] Courtesy NASA Photo S-69-47905.

5.1 Apollo Scientific Results

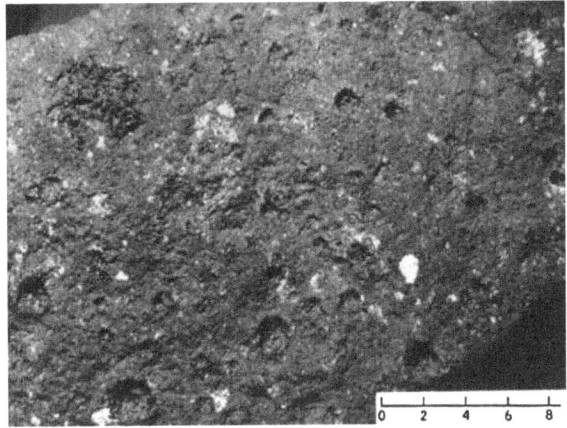

Fig. 5.2 Micrometeoroid impacts (Zap pits) on 10019. Scale mms

bombardment over the aeons. So while it was clear from unmanned missions and again from Apollo 11 that the lunar surface was safe for landing, the lunar surface was indeed composed of dust, rubble, and other products of impacts in the regolith. At the Apollo 11 site its depth varied from about 3–6 m.

The composition and petrology of the Mare basalts was similar to terrestrial basalts with minerals familiar to every petrologist, such as pyroxene, plagioclase, olivine, but with the important exception that they had been formed in a highly reducing and very dry system. This resulted in the occurrence of native iron and troilite, not seen in typical terrestrial basalts. Among the Apollo 11 basalts two groups were recognized, one had low Potassium, which was slightly older, and the other high Potassium. These results confirmed those from the landers and finally laid to rest any ideas that the moon was a primitive body. Rather it had to be an evolved body that had a crust and a deeper source for the mare basalts.

The Apollo 11 breccia samples were all products of relatively small impacts in the local lunar regolith. This was pretty much the same for the Apollo 12 site, which was again a mare site. However, with later missions sampling the highlands, very different breccias were sampled. The Apollo 14 on the Fra Mauro formation and Apollo 16 on the Cayley formation were on ejecta from massive basin forming events. Apollo 15 was close enough to the Apennine Mountains to have material from them on the site. Finally at the Apollo 17 site in the Taurus Littrow region, boulders that had rolled down from the massifs surrounding the site would provide a chance of sampling sequences of rocks from the highlands.

The early efforts focused on the mare basalts, which were accessible and provided detailed history from about 3.2 to 3.7 Ga. However, by mass they were a small fraction of the highlands, which would be key to the earlier history. The primary highland rocks were also igneous, as geologist would say, having cooled from magma. The difference between them and the lavas was that they had cooled at depth within the lunar crust rather than on the surface, and so cooling had been much slower and larger coarse grained crystals had time to form. The highland sites also allowed sampling of the ejecta blankets from giant basin forming events.

Fig. 5.3 Lunar seismograms: LPX, LPY, LPZ, long period seismograms X, Y, and Z components, SPZ short period vertical component

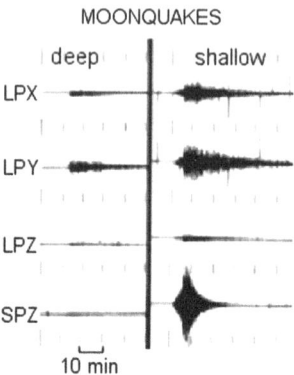

These ejecta will play a major part in the paleomagnetic story and will be covered in more detail below, when we discuss that record.

The passive seismometer recorded a number of events and the similarity of the signal between the natural events and the impact of the Apollo 11 LM led to their interpretation as either moonquakes, or impacts (Fig. 5.3).[4] These signals were quite unlike those we were familiar with from seismology on earth, with their distinct P waves, S waves and surface waves that Oldham had recognized. The lunar seismograms are consistent with passage through material with low loss, or good transmission quality, but also strong scattering. They did not look very promising for telling us about the deep structure of the moon, but they did show that the moon was tectonically much quieter than the earth.

One particularly prescient paper in the Apollo 11 Science volume came from John Wood and colleagues on the Lunar Anorthosites, they had seen that a few percent of the soil fragments in the range from 1 to 5 mm were quite different from the basalts and breccias, which formed the bulk of the particles, such as we had seen in our soil samples. They were light colored in contrast to the basalts and had low densities. On earth similar rocks are found towards the top of large slab like layered intrusives. The bodies are compositionally different at the top and bottom. Skipping over many details, the basic idea is that as crystallization proceeds the denser crystals sink and the lower density crystals rise. The authors suggested that the highland crust was dominantly composed of these lower density light colored rocks. This idea later became the basis for the famous magma ocean model (Fig. 5.4),[5,6] according to which the lower density original crust floated on the

---

[4] Courtesy Science at NASA. See also The lunar seismic network: Mission update, Neal. C.R., et al., 2004, Lunar and Planetary Science Conference, XXXV, Abstract.

[5] Figure from Wikipedia. See also Taylor, G,J,. 1994, *The Scientific Legacy of Apollo*. Scientific American, 270, 6, 40–47. In this article there is a very useful summary of knowledge of the moon in a number of key areas before and after Apollo. The author also reminds us that ideas we will meet in the last chapter in this book on the origin of the moon, were anticipated by Reginald Daly in 1946.

[6] NASA Apollo 16 mission gallery.

5.1 Apollo Scientific Results

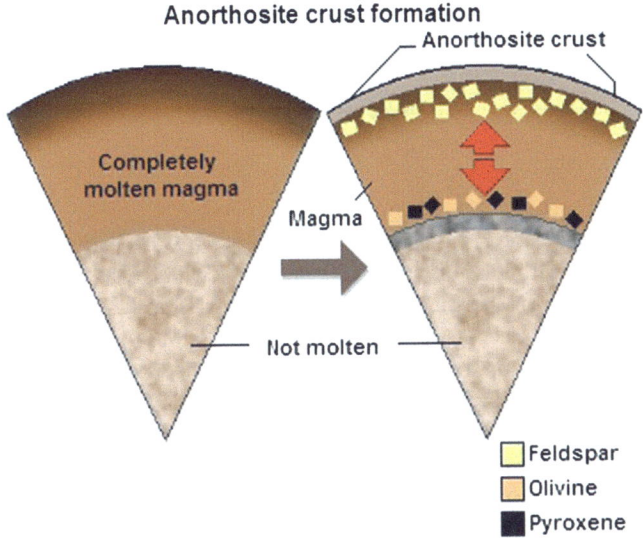

Fig. 5.4 The lunar magma ocean yielding anorthosite crust above and more basic residual below

higher density material. This eventually gave a lunar anorthositic crust of some 25 kms, standing higher than the Mare that were derived later from the higher density residual below.

The samples carried a record of shock, as was noted in several papers at the Apollo 11 meeting. The basalts were for the most part relatively weakly shocked compared with the breccias and many of the individual soil grains. It was hardly surprising that the samples showed petrological effects of shock since the moon has been bombarded for billions of years and each event must have generated shock effects. Locally shock will generate heat and as the level of shock increases shock melting and eventually vaporization is reached. This can happen on the small or large scale. It is of course of interest as a phenomenon of its own, but it is critically important to us because it can affect magnetization, and even in the right circumstances cause magnetization.

5.2 Lunar Stratigraphy

Lunar stratigraphy records the history of events on the moon. On earth one of the fundamental achievements of geologists in the 18th and 19th centuries was the division of the succession of rocks into geological systems, corresponding to

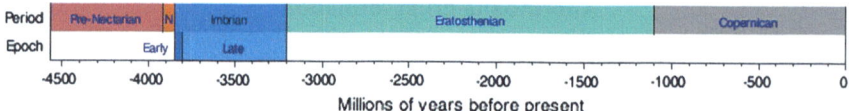

Fig. 5.5 Lunar stratigraphy

geological time units, or periods. William Smith (1769–1839) was an engineer involved in the construction of the canal network, which was the major heavy transportation system of the time in England. He saw that fossils found in rocks occurred in systematic successions, so that it was possible to correlate rock units on the fossils present and build up the history of the rocks according to this principle of faunal and floral succession. He published the first geologic map of England and Wales early in the 19th century ($\sim$1815). Much later it was possible to add absolute time markers from radioactive age determination.

The situation on the moon was quite different because there were no fossils to use. However, cratering had been going on throughout lunar history, so that in principle the relative age of units should control their crater density—the longer a unit had been exposed the greater the crater count. A major effort in crater counting was underway before Apollo and continued on through the Apollo missions and to subsequent unmanned missions. As on earth absolute time markers could be used to give absolute ages to these relative age units. Figure 5.5[7] gives absolute ages for the lunar time periods of Pre-Nectarian, Nectarian, Early and Late Imbrian, Eratosthenian and Copernican. The names Nectarian, Imbrium and Eratosthenian arise because there are extensive deposits from these major basins, which serve as key stratigraphic markers and define the rock systems and time periods with which we will be principally concerned.

5.3 Paleomagnetism of Apollo Samples

Finally and most importantly from our viewpoint the Apollo 11 sample collection was surveyed magnetically by Dick Doell and Sherm Gromme and found to be magnetized. To many this was a surprise because the Russian satellite had shown that the moon did not have a magnetic field like the earth and that any surface field could not be greater than about a microTesla (μT). Again we are encountering the units that can give so many problems, but we will again use comparative

---

[7] This convenient figure comes from "Lunar time scale" Wikipedia. For additional coverage Heiken G, Vaniman D, French BM (1991) *Lunar sourcebook: a user's guide to the moon*, Lunar Science Institute, Cambridge University Press. Published just over 20 years after the Apollo 11 landings is the essential summary of early work and was followed by the updated comprehensive text *New Views of the Moon* eds. Joliff Bl, Wieczorek MA, Shearer CK, Neal CR (2006) Rev. Mineralogy and Geochemistry, 60, pages 721.

5.3 Paleomagnetism of Apollo Samples

illustrations. The result from the Russian satellite told us that the surface field on the moon was about 50 times smaller than the surface field on the earth. This result was confirmed by Explorer 33. Yet the samples carried remanent magnetization, a memory of an earlier magnetizing field on the moon much stronger than the present weak field.

Let's immediately clarify some of these ideas in magnetism that we will need from here on. Bearing in mind, that a lot of technical details is not everyone's cup of tea, I will try to minimize them and banish more technical parts to appendices, but let us at least be clear about remanant magnetization, or as it is sometimes called magnetic remanence, which is the star of our story.

Our starting point is the familiar magnetic hysteresis loop (Fig. 5.6), which you probably saw in high school. The sample starts with zero magnetization in zero magnetic field at the center of the figure. A magnetic field is applied with an electromagnet and we measure the magnetism of the sample with a special magnetometer, which works in the presence of strong magnetic fields. The field is first increased to take the sample to saturation magnetization. No matter how much more we increase the field, the magnetization will not increase any further—it has saturated. After this, the field is decreased to zero. At this point even with zero magnetic fields, the magnetization has not decreased to zero, but there is a positive Remanent Magnetization or Magnetic Remanence—a memory of the positive field. A negative field is next applied and eventually the magnetization falls to zero. This field that brings the magnetization to zero is called the coercive force field. It is a measure of how strongly the remanent magnetization is held by the sample. Continuing on from there a symmetrical part of the loop takes the sample to negative saturation magnetization, then as the negative field is reduced the sample reaches the negative magnetic remanence at zero field, with increasing positive field the coercive force is reached and finally the magnetization returns to positive saturation magnetization.

It is worth noting here that magnetic particles can be usefully divided into two categories: single domain and multidomain particles. This had been recognized

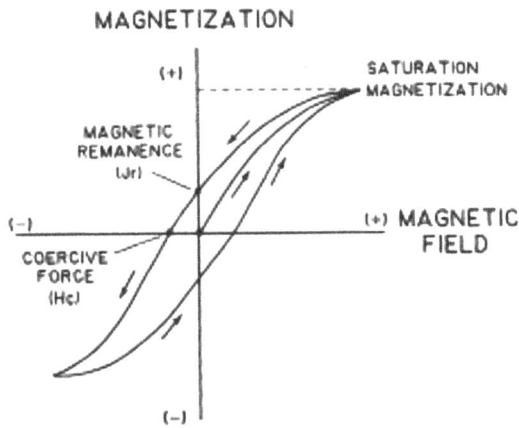

Fig. 5.6 Magnetic hysteresis loop

long ago. Single domain particles are small, for example in iron they are fractions of a micron. That is hundreds of times smaller than the thickness of a human hair. They have remanent magnetization that is pretty much homogeneous throughout the particle. Their hysteresis loops tend to be fat like the schematic one in Fig. 5.6. The remanence tends to be strongly fixed in the particle, so that a high field is needed to change the magnetization. In large grains, the magnetization in the particle is divided up into mutually opposing regions, or magnetic domains, so that the energy expended by the particle in its external field is reduced. This energy depends on the net remanent magnetization of the particle, so that by having self compensating magnetization in the particle its external field is much reduced. Changes in the magnetization of these particles are achieved by motion of the walls defining the magnetic domains. This does not require strong fields. In complete contrast to the hysteresis loops of single domain distributions the hysteresis loops of multidomain samples are skinny, as we found with sample 14053.48, an older mare basalt. Remanent magnetization in multidomain particles is very easy to reset in weak fields.

To paleomagnetists the single domain particles are our friends and carry the best record, whereas the multidomain particles are the bad guys that make our work harder because we have to see through the noise they tend to generate. It appeared that there were lunar samples whose magnetic properties were dominated by single domain iron but they were few and in the majority the iron was strongly multidomain. Evidently some samples were likely to be good recorders, but others would be difficult for us to study.

The moon does not have a magnetic field like the earth, so how could these lunar samples have been magnetized. Was their magnetism a record of an earlier time when there was a field, or was it an artifact. Possibly the samples acquired their magnetization during collection, or from fields in the spacecraft, or during sample preparation.

5.4 Magnetism of Lunar Crust: Surface and Sub-satellite Observations

Other observations soon became available that showed there was genuine lunar magnetism. On Apollo 12, 14, 15, and 16, magnetometers were taken to the moon and used to measure the surface magnetic fields at the various sites in experiments designed by Dyal, Parkin and Sonett. At all sites, with the possible exception of Apollo 15, the magnetic fields were larger than the limit set by the Russian and Explorer 33 results. The surface fields at the mare sites were noticeably lower than at the highland sites (Fig. 5.7).

At the Apollo 16 site several measurements were made with a portable magnetometer and the results showed that the field varied in direction and strength to such an extent that local sources were required to explain the observations. Now in

5.4 Magnetism of Lunar Crust: Surface and Sub-satellite Observations

Fig. 5.7 Apollo 16 magnetometers: **a** surface and **b** portable

addition to the samples, the lunar surface evidently exhibited remanent magnetization.

The third observation of lunar magnetism came from the sub-satellites launched from the command module in orbit around the moon during the Apollo 15 and 16 missions, which had magnetometers on board. It soon became clear that the Mare were relatively bland magnetically with no strong features, whereas the stronger anomalies were in the highland areas and the strongest of all were on the far side of the moon, which is notably free of mare.

At this point serendipidity played an important role. The sub-satellites had onboard an experiment to measure electrons in the vicinity of the moon. The experimenters, Prof. Anderson and Bob Lin of UC Berkeley, were surprised to find electrons apparently reflected from the lunar surface. This was not supposed to happen. The electrons should have been absorbed by the lunar surface. However, this Berkeley group was smart enough to realize why electrons might be reflected from the lunar surface and it turned out that these fields were defined with very high resolution by this method. The explanation involves the same ideas that we encountered in thinking about the Van Allen radiation belts in the Earth's magnetosphere. Remember that an electron traveling in a magnetic field, experiences the Lorentz force, which makes it turn and spiral around the line of force. If the field gets stronger, the pitch angle will steepen. Eventually the electron will be reflected back out of the region of strong field, just as the charged particles in the earth's radiation belts are near to the poles. Evidently, there must have been strong magnetic fields near the moon's surface that were reflecting the electrons.

These observations then revealed a number of important results. They found local high fields, which turned out to be associated with lighter regions on the lunar surface. Initially this was something of a mystery too. However, the moon's surface darkens with time as it is bombarded with radiation principally from the solar wind, but for some reasons these swirls had escaped and were all light. The light regions might be light because they were young, as are the rays form young craters. If they were not young, the regions must have been somehow protected from the radiation that darkens the surface. In this case, it was known that the regions were of a various ages, but were associated with strong fields. The strong magnetic fields had evidently reflected particles in the solar wind, so that the surface had been protected from the darkening processes. The next surprise was that the light areas appeared to be preferentially located antipodal to the younger large basins, whose ages were about 3.6–3.9 Ga. We will leave the discussion of these fields until later and simply note that this was yet another indication of the magnetization of the lunar crust.

5.5 Summary

The results discussed at the Apollo 11 Science conference began to make clear how much we had learnt about the moon. In particular, the division of the lunar crust into highland and mare regions seemed to be satisfactorily explained by the magma ocean model. According to this idea, the highland crust was the initial low density crystallization product, whereas the basalts of the mare regions were later derivatives of the denser crystallization product that had sunk in the original magma ocean. More immediately relevant to our problem of lunar magnetism, it was clear that the lunar samples carried remanent magnetization. By the time, the surface magnetometers had recorded local remanent magnetic fields on the lunar surface, the orbiting satellite magnetometers and the electron reflectance experiment had shown large scale magnetic anomalies, it was obvious that the magnetism of the samples was not likely to be due to artifacts resulting from their collection, transportation and treatment on earth. As we have already seen, terrestrial rocks acquire remanent magnetization, called Natural Remanent Magnetization when they are formed in the geomagnetic field, so next we will look at terrestrial remanent magnetization and see how it might help understand the magnetism of lunar samples.

Chapter 6
The Earth's Magnetism: Paleomagnetism as a Rosetta Stone for Earth History

Paleomagnetism, the record of the geomagnetic field carried by rocks, has played a key role in our understanding of Earth. In this chapter, we review the history of our understanding of the Earth's magnetic field and its paleomagnetic record, as an indication of what the possible record of a lunar field might yield. We also begin to consider the ubiquitous magnetic fields in the cosmos and how they are generated and maintained.

6.1 The Geomagnetic Field and Its Paleomagnetic Record

The development of our understanding of the geomagnetic field is a fascinating story and central to understanding the magnetism of other rocky planetary bodies and moons. A discussion of geomagnetism should probably start with the Chinese. Joseph Needham suggests in his great text "Science and Civilization in China" that the development of the compass may have been China's greatest gift to humanity. It was made of the loving stone as the Chinese referred to lodestone, and in the form of a spoon on a bronze plate, so that the spoon was free to rotate and show the field direction. It was used in geomancy, whereby houses, beds and the like were aligned with the local geomagnetic field, which was held to be beneficial. The Chinese described this as a south-seeking instrument. In both China and in the west magnets were used in their key role as an essential aid to navigation at sea.[1,2]

---

[1] When I was a student at Caius College Cambridge, Joseph Needham was writing his great text and I remember going to lectures of his. It was a joke among some of us that since each volume got larger than the last it was not clear that the task would ever be finished. However, it was and became one of the great scholarly works of the mid-20th Century. The references to magnetism are in Volume 4, part 1 Needham (1962).

[2] Aczell gives a delightful account of the early development of the compass in the west, leading to the boxed version with its compass card produced in Amalfi between 1295 and 1302. By this time, Amalfi already had a long history as a maritime power. When it was replaced by the states of Genoa and Venice, the compass continued to play a major role, as it has done subsequently for all seafarers (Aczell 2001).

As late as the 12th century, one encounters in Europe the idea that the alignment of the compass arises from its attempt to follow the pole star. However, with the work of Petrus Peregrinus (Peter the Pilgrim) in the 13th century, the scientific study of the magnet can be said to begin in the west. He had used small magnetic needles placed on the surface of a sphere of lodestone to study its magnetic field. He had recognized the poles, where the needles stood on end, and the magnetic equator, where they lay parallel to the surface Fig. 6.1[3]. Essentially, he had established the major features of the magnetic field of a sphere, a dipole field. Such a field is produced by a current loop, or small magnet. It is similar to the field revealed by iron filings over a bar magnet. Petrus Peregrinus worked in the high middle ages, was a contemporary of St. Thomas Aquinas, living at a time when the great cathedrals of the Middle Ages were being built.

William Gilbert, the next giant in the history of the geomagnetic field was born in Colchester in 1540. He attended St. John's College, Cambridge. In his twenties, he traveled widely on the continent. In 1573, he was elected to the Royal College of Physicians. Later, Queen Elizabeth appointed him as her physician in-ordinary. His early scientific investigations were in chemistry, but they soon gave way to his life's work in electricity and magnetism. His great achievement was to make the leap of imagination to see the analogy between the known field of the lodestone sphere, as demonstrated by Petrus Peregrinus, and the field of the Earth: "Magnus magnes est ipse Globus terrestris". This took place about 300 years after the original work by Petrus Peregrinus. Gilbert's treatise on the magnet is one of the first modern experimental discourses. De Magnete was recognized by no less than Galileo and Kepler to be of major importance: in his time Gilbert was revered as Newton and Einstein would later be. Although his ideas were a curious mixture of science and mysticism, he saw before Francis Bacon the power of the experimental method. He had a profound effect of the world of his time and was an important part of the flowering of talent in the Tudor England of the first Queen Elizabeth, the daughter of Henry VIII and Ann Boleyn. By Elizabethan times, the compass was in common use for navigation.

Declination was known in China from at least about 500 AD, in Tang dynasty times and maybe much earlier. It was probably first recognized in Europe by the pocket sundial makers of Nuremberg. Certainly by the end of the 16th century, Mercator had seen that it was the principal cause of error in contemporary map making. The discovery of the inclination of the geomagnetic field vector is usually credited to Norman. However, some 40 years earlier, in 1544, a certain vicar of St. Sebald's in Nuremberg, George Haartmen, had discovered that the field vector was inclined from the horizontal and had written to his superior, to tell him of the discovery. Unfortunately, the Duke was not impressed with the curious interests of the vicar and the letter lay unknown to the world until 1831. The discovery that the inclination and declination of the geomagnetic field changed with time was made

---

[3] Private photograph taken, I believe, by Vic Schmidt, who joined our research group from Carnegie Tech., as it was then known.

6.1 The Geomagnetic Field and Its Paleomagnetic Record

by Henry Gellibrand. This showed that the earth's magnetic field was dynamic, changing with time and not a static fixed phenomenon. Edmund Halley suggested that it was due to differences in the rotation rates of the crust and the deeper source of the magnetic field.

While the time dependence of the field was being established, so too was the variation of the field over the surface of the Earth. In 1700, Halley made an oceanographic survey, the prime purpose of which was to study the field over the ocean surfaces. A century later, Von Humboldt measured the relative intensities of the horizontal field in South America and Europe. This work was brought to culmination by the genius of Carl Freidrich Gauss, who developed a method for measuring the intensity of the magnetic field and for its mathematical analysis. His analysis allowed him to separate the field into internal and external parts. He then proved that to the accuracy of his observations the Earth's magnetic field was of internal origin. Thus roughly another 300 years after Gilbert had recognized that the field was of internal origin, his idea was proved to be correct, and so the major steps in our understanding of the geomagnetic field in western civilization had taken place with roughly 300 year intervals.

Whether one speaks of a north-seeking compass as we do now, or a south-seeking compass as the Chinese did, has no fundamental significance. However, this is perhaps a good place to clarify one point that sometimes causes confusion. In speaking of a compass, we refer to it seeking geographic north. As the north magnetic pole of a compass must seek a south magnetic pole, there must then be a magnetic south pole near the north geographic pole. The north geographic pole is indeed a south magnetic pole.

In Fig. 6.1[3] we showed the configuration of the surface field of the lodestone sphere. Following this simple pattern of what is termed a magnetic dipole field (the field of a pair of north and south magnetic poles), we can predict the variation of the geomagnetic field over the surface of the earth. For example, if the field were caused by a centered, axial dipole, the lines of equal inclination would everywhere be parallel to the lines of latitude and the declination would be zero everywhere. However, in reality the lines of equal inclination depart from the lines of latitude. Moreover, the poles of the magnetic field also depart from the geographic poles. Evidently, the field cannot be due to a simple centered dipole (Fig. 6.2). The principal features of the inclination of the earth's field show that the dipole must depart from alignment with the rotation axis. It is actually inclined by some 12°. If we subtract the dipole mathematically from the total field, we find that there remain features of different wavelengths, which follow a distinctive pattern. There are long wavelength features due to core fields with long-range coherence across the major areas of surface of the Earth, and shorter wavelength features which are from fields due to the crustal magnetic anomalies, which extend over much shorter distances. These crustal anomalies are the terrestrial analogues of the lunar magnetic anomalies seen with the sub-satellite magnetometers that we will discuss later. The moon of course now lacks the equivalent of the geomagnetic dipole.

The changes with time of the field, discovered by Gellibrand are termed secular variation. The highest frequency changes of the geomagnetic field are due to the

Fig. 6.1 A modern version of the experiment of Petrus Peregrinus

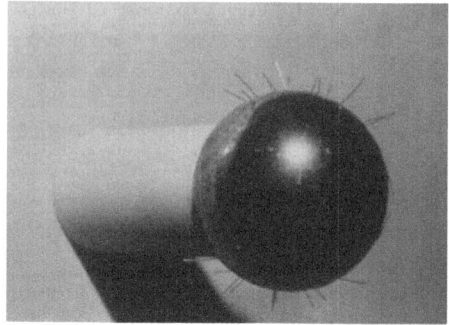

external field, but at periods of tens of years and beyond this contribution becomes insignificant. Conversely, rapid fluctuations due to the internal field cannot penetrate the finite electrical conductivity of the mantle, if they have periods of less than about 10 years and so fields due to internal changes predominantly have periods of hundreds to thousands of years and longer.

The most profound change in the geomagnetic field is of course its reversal, when the north and south poles swap their geographic locations. This is one of the most important results from paleomagnetism and was reported by Brunhes and his colleague David early in the 20th century. The French and others had built up an impressive record of field changes on an archaeological time scale by using pottery. As the pottery cools after the firing of the clay, it acquires a remanent magnetization and in so doing records the ambient field. It appears that Brunhes had a friend in the part of the French government dealing with road construction. Knowing the success of the baked clays used in pottery as recorders of the magnetic field, Brunhes asked his friend to let him know, if the road crews came across any naturally baked clays, such as one finds beneath lavas or alongside dykes. The Gods of paleomagnetism smiled and Brunhes's friend did come across such a situation. When Brunhes studied the rocks he got a big shock. The rocks were magnetized just oppositely to the local field in France at that locality. The remarkable discovery of Brunhes and David is clearly one of the most important aspects of the earth's magnetic field. The rocks were of course much older than the archaeological samples and eventually it turned out that the last reversal was some 780,000 years ago.

We paleomagnetists spent a lot of time arguing about the reality of reversals and testing other possibilities to explain the observation, but the work of Cox, Doell and Dalrymple of Stanford and the U.S. Geological Survey at Menlo Park and of McDougal and Tarling in Australia, clarified the issue by showing that there was a succession of reversals recorded by the rocks at intervals of some hundred thousands of years for the past few million years. We then tried to find successions of samples that were formed when a reversal was taking place to see what happened. This could be an important clue to how the geomagnetic field works.

One of the most exciting pieces of research by our group was tracking down a reversal recorded in the Tatoosh intrusion in Mount Rainier National Park,

6.1 The Geomagnetic Field and Its Paleomagnetic Record

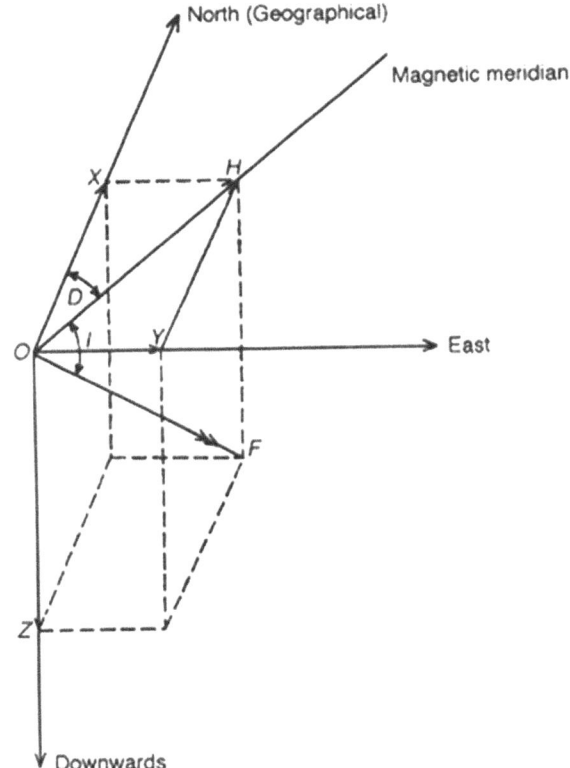

Fig. 6.2 Illustration of declination (*D*) and inclination (*I*)

although I am not sure that it has ever been generally accepted as a reliable record by our paleomagnetic colleagues. From this record, it appeared that the field first decreases in strength substantially, maybe to about 10 %. It also loses the comparative simplicity of its pattern on the earth's surface, with the secular variation relatively enhanced, as the main dipole seemed to weaken. After a period of low field strength, recovery gives a return to a simpler, but reversed pattern with growth of the dipole field and a return to weaker secular variation. The time scale of the intensity changes seems to be several thousand years and is longer than the directional changes. Similar results have been found in other records.

Remembering that the geomagnetic field holds off the solar wind and protects the earth from harmful radiation, the reversal could mean a bad time on an earth. However, reversals have taken place about every few hundred thousand years for millions of years, so life has obviously survived many reversals.

We may possibly be approaching a reversal in the next few thousand years—we are certainly overdue for one and the field appears to be showing some of the behavior that we would expect before a reversal. The effect of a reversal of the geomagnetic field could be confined to extending aurora to all latitudes and a minor increase of cancer incidence, which may be eliminated by then anyway. Yet, the effects of reversals on an advanced technological society with vulnerable power grids could be a good deal more serious than for Homo erectus and whoever else was around 780,000 years ago. Fortunately, we should have plenty of warning.

6.2 Magnetic Fields of the Cosmos

Is our geomagnetic field unique, or do we see magnetic fields elsewhere in the cosmos? Actually the cosmos abounds with magnetic fields. From the scale of the atom, to the galaxy itself, magnetic fields are everywhere. Indeed the very thoughts in our brains and the beating of our hearts have magnetic fields associated with them. The Sun has a magnetic field, which also reverses. It gives rise to an interplanetary field, which permeates the solar system. The giant outer planets—Jupiter, Saturn, Uranus, and Neptune—all have magnetic fields. Among the inner planets, the Earth, and probably Mercury have magnetic fields. Mars almost certainly had one, but does not now. In the Sun, the giant outer planets, and the Earth, the fields are maintained by magnetohydrodynamic processes taking place in regions of high electrical conductivity. The Sun is not alone among the stars in having a magnetic field—far stronger fields are common.

What causes these universal magnetic fields? The key mechanism is the motion of electric charges, as has been known since the time of Hans Christian Oersted, who noticed, apparently during a lecture he was giving, that when he brought a compass near to a wire carrying an electric current the compass was deflected. However, the generation of the ubiquitous magnetic fields of the cosmos is not a matter of electric wires threading the universe, but a more subtle mechanism whereby magnetic fields are amplified from seed fields. On the cosmic scale these seed fields may take the form of turbulent flow of ionized material with their charged particles, although discussions of their origin tend to be a bit vague.

Let us assume that seed fields are available, then next we need to recognize that most of the universe is a very good electrical conductor, unlike the atmosphere of our planet, which is a very poor electrical conductor. Given a large scale electrically conducting medium with internal motions, magnetic fields will follow, pretty much as night follows day. The field lines are distorted and stirred up by the motion of the conductor in which they are trapped. As a result of the work of a number of theoreticians at the middle of the last century led by Walter Elsasser and Sir Edward Bullard, mechanisms were proposed whereby the geomagnetic field can be maintained by processes going on in the electrically conducting fluid of the outer core. The details of this process get complicated and so are addressed in Appendix 5. We just need to remember that the generation of magnetic fields in highly conducting media by the mechanical energy of their internal motions is universal.

6.3 The Paleomagnetic Record as a Rosetta Stone for Earth History

The reason for calling the record of Earth's magnetic field a Rosetta Stone for Earth history is simply that it has provided a key to fundamental aspects of earth's story. Initially, it solved the long standing problem of continental drift, which set geology on the right track leading to Plate Tectonic theory—the fundamental unifying theory of geology.[4] It was the record of reversals of the geomagnetic field in the sea floor magnetic anomalies that convincingly demonstrated that new ocean crust was formed at the ocean ridges in the process that became known as sea floor spreading. Fred Vine and Drum Matthews from the Geophysics department at Cambridge University saw this and published a famous paper in Nature. Sadly, in one of those all too frequent injustices of science, the Canadian Morley,[5] who had come up with the same idea independently and almost simultaneously with Vine and Matthews did not initially receive the credit for his idea because he had encountered bad reviewers, who were not smart enough to recognize how good his idea was. A paper, which outlined the basis for one of the most important ideas in the earth sciences was rejected by Nature and the Journal of Geophysical Research (JGR), but fortunately he presented is ideas at a meeting of the Royal Society of Canada in 1964, so that his contribution was eventually recognized.

The reversal chronology recorded by the sea floor anomalies and over a longer period via the ocean sediment records, and the paleomagnetic reversal stratigraphy of basalts, when combined with key absolute ages gave a time scale for earth history with unmatched resolution that could be used for studies of climate changes, paleontology and anthropology. What made all of the results that we described above possible was the remarkable and fortuitous phenomena by which rocks record the fields in which they were formed. Remanent magnetization is not only acquired by exposure to magnetic field as in hysteresis loops, but by the process formation of many rocks. It is now time to explain how rocks acquire their remanent magnetization and become Rosetta stones.

6.4 Paleomagnetism: How the Magnetic Field is Recorded by Rocks

Most rocks on earth contain iron oxides, such as magnetite or hematite and sulphides pyrrhotite or greigite. Fine particles of these minerals are excellent magnetic memory elements. Once magnetized, the fine particle magnets found in

---

[4] Excellent coverage of the story of the convergence of paleomagnetism and age determination leading to sea floor spreading and plate tectonics (Glen 1982).
[5] Morley's recognition of the seafloor imprinting of the reversal is recounted in fascinating detail in page 301, 302 of Glen's book. It is worth noting that Bob Dietz was in the forefront here on sea floor spreading, as he was in the interpretation of craters as impact features.

rocks need strong fields, or some other energetic process to reset their magnetization. Given this, the reader may well ask how the weak geomagnetic field manages to set their magnetism in the first place.[6]

The simplest type of remanent magnetism to understand is that of sediments on earth, called depositional remanent magnetization (DRM). I apologize for the confusing alphabet soup of mechanisms of magnetization that we paleomagnetist have developed, although we are hardly alone in this failing.[7] Try visiting a hospital, or NASA! I will avoid this soup as much as possible. To return, when sediments accumulate on the sea floor, or elsewhere, the magnetic particles they contain become preferentially aligned parallel to the geomagnetic field, just as does a compass needle. The alignment of the particles in the sediment is however, far from perfect and so depositional remanence in the geomagnetic field is about ~1 part in 1,000 of the maximum remanent magnetization the sediment would carry if it were exposed to a saturation field, as in the production of the hysteresis loop. The general term for the remanent magnetization found in a rock is the natural remanent magnetization (NRM). Depositional remanence is not likely to play a key role in lunar rocks.

Probably the most important type of remanent magnetization in terrestrial and lunar samples is that acquired when rocks cool in a magnetic field. It is called thermal remanent magnetization (TRM). It is how the samples Brunhes studied and how the great pile of lavas outside of my window here in Oahu got their magnetization. This thermal remanence mechanism is sometimes misrepresented as the alignment of particles in molten magma that later becomes a rock. Actually, TRM is usually acquired many 100s of degrees below the temperature at which igneous rock solidifies. The properties of TRM were established in the 1930s and 1940s by three giants in our field-Koenigsberger in Germany, Thellier in France and Nagata in Japan.

As we heat a rock to high temperature its magnetic minerals eventually reach their Curie point—the temperature at which they lose their magnetism (Appendix 1). A standard plot of this loss of saturation magnetization with temperature is given for a lunar sample in (Fig. 6.3), (Nagata et al. 1972). On the left hand side at lower temperature, the exchange force, which makes the iron magnetic, dominates. As the sample is heated, magnetization decreases, until at the Curie point the iron ceases to be magnetic. Thermal energy has overcome the exchange energy.

As we noted in the Appendix 1, there is competition in magnetic materials between an ordering force, the exchange force giving magnetism, against the disordering effect of thermal energy. When a rock like a lunar sample carrying magnetic minerals cools through its Curie point and on down to room temperature

---

[6] There are by now excellent texts covering the magnetism of rocks and the record they carry, The most comprehensive and the standard text of Rock Magnetism is by Dunlop and Ozdemir (1997). For an account, which includes more paleomagnetism and geological applications is another excellent source (Butler 1992).

[7] The alphabet soup of mechanisms of remanent magnetizations is a constant source of irritation to those not in the paleomagnetism club and so I have included a glossary (Appendix 3).

6.4 Paleomagnetism: How the Magnetic Field is Recorded by Rocks

in the presence of a magnetic field, it acquires thermal remanent magnetization (TRM). A beautiful theory of TRM was presented by Néel, which has become a cornerstone of our understanding of rock magnetism. At the risk of going too far into details, I'll see if I can give an impression of this theory in Appendix 6. An important feature of this theory is the recognition of the blocking temperature at which the remanent magnetization is fixed in the magnetic particle as TRM. Meanwhile, we will utilize this remanent magnetization acquired when lunar samples cooled on the moon for our record of lunar fields. TRM in the geomagnetic field is about 1 part in 100 of the saturation remanent magnetization of the hysteresis loop.

In a somewhat analogous way to the acquisition of TRM as temperature falls, when particles of magnetic material grow in the presence of a magnetic field, they pass from a stage in which they are too small to carry stable remanent magnetization to a critical size at which magnetization stabilizes, overcoming thermal energy and recording the ambient field. Like TRM in the earth's field, it is about one part in 100 of saturation remanent magnetization. This chemical, or crystallization remanent magnetization (CRM) is likely to be a source of NRM on the moon.

Several other types of NRM are also likely to be found on the moon and we will hear a lot about shock remanent magnetization (SRM), which is self evidently acquired when a rock is shocked in the presence of a field. Viscous remanent magnetization (VRM) is another important remanent magnetization for us. It is carried by particles close in size to the border between stable and unstable behavior. This VRM is retained over a range of times depending upon the temperature and grain size of the particles. As we shall see, it can be a source of contamination, when lunar samples are stored on earth in the geomagnetic field.

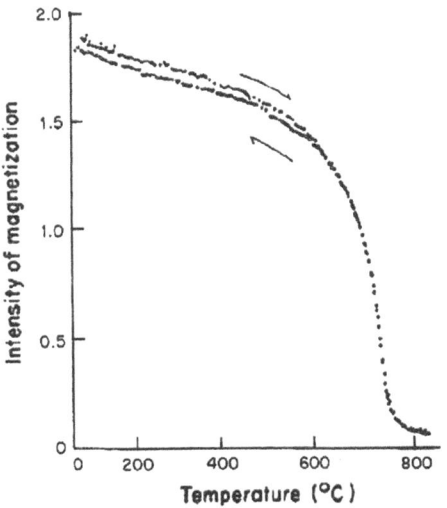

Fig. 6.3 Thermomagnetic curve as a lunar sample is heated

6.5 Summary

The history of the geomagnetic field as recorded by paleomagnetism has illuminated much earth history. We take advantage of this record preserved mostly by thermal remanent magnetization. We can collect samples on earth whose magnetization was acquired millions, or even billions of years ago, measure it, and build up the history of the geomagnetic field. Now the question was could the same mechanisms in the lunar mare or breccias record the lunar fields and provide a Rosetta Stone for lunar history. Was an ancient lunar field a simple dipole field like earth's roughly aligned with the present rotation axis? Were there reversals of that field?

References

Aczell AD (2001) The riddle of the compass, Harcourt Press, California
Butler RF (1992) From magnetic domains to geologic terranes. Blackwell Scientific Publications, New Jersey
Dunlop DJ, Ozdemir O (1997) Rock magnetism fundamentals and frontiers, Cambridge University Press, Cambridge
Glen W (1982) The road to Jaramillo, Stanford University Press, California
Nagata T, Fisher RM, Schwerer FC (1972) Lunar rock magnetism. The Moon 4:160–186
Needham J (1962) Science and civilization in China, Cambridge University Press, Cambridge

Chapter 7
Lunar Paleomagnetism: Methods and Preliminaries

The lunar samples were collected from the regolith, having been brought there by impacts that excavated them from their previous locations. The orientation in which they originally formed is therefore unknown. Thus tests of the reliability of their record depending upon planetary wide consistencies of paleomagnetic directions cannot be used, as they are on earth. Our approach has been to use internal consistencies within samples and demagnetization characteristics, which serve as fingerprints for particular types of remanent magnetization.

7.1 The Continuing Debate Over Lunar Magnetism

More than 40 years after the Apollo[1] days, the case for a lunar dynamo remained unconvincing for many and a famous paper by Kristin Lawrence and her colleagues at UC San Diego came out suggesting that the case for an early lunar dynamo was uneasy (Lawrence et al. 2008). At the same time very interesting positive evidence for the early dynamo was coming from the MIT group led by Ben Weiss (Garrick-Bethell et al. 2009). In addition the French group from Aix-en-Provence, consisting of, Cécile Cournède Jérôme Gattacceca and Pierre Rochette, documented additional samples that appeared to be carrying primary NRM recording an early dynamo. They also provided an ingenious approach for testing for a centered lunar dipole. Their assumption was that the magnetic fabric or alignment of planar particles, was the ancient horizontal. If the inclinations of the magnetization to this horizontal from various sites gave consistent values for a dipole field then this was good test and they did get a positive result (Cournède et al. 2012). To test this properly we need oriented samples from bedrock, which are not likely to be available any time soon. At the moment we are in the middle of a renewed effort to test the idea of lunar magnetism rigorously, so let's see where we stand.

---

[1] Apollo 17 photograph AS17-134-20425. Courtesy NASA.

Fig. 7.1 Astronaut Harrison Schmitt collecting a sample from the lunar regolith

The samples were collected from the regolith (Fig. 7.1) and in some cases the orientation on the lunar surface in which samples were found was recorded. Our task was to see if any NRM had remained from when the rocks were formed billions of years ago and what that NRM might tell us about any lunar magnetic fields present then.

Conceivably the NRM might have been picked up in the process that brought them into the regolith, during time on the lunar surface, in the spacecraft, or even during handling on earth. I remember Keith Runcorn saying to me once that I seemed to have considered all possible mechanisms for the origin of the NRM, except that the samples had been touched by the hand of God. With his customary good sense, he had soon recognized that the NRM was most likely a record of an early lunar field. However, we were all conscious that it would be nice to get the interpretation of the moon's magnetism right first time. Several research groups around the world were involved and it was a joy to work with so many colleagues, most of whom were old friends. Dave Strangway from the University of Toronto took the lead spending time at the Space Center as a coordinator of the efforts and of course Keith Runcorn was in the vanguard.

Over the years, a number of field tests had been developed in terrestrial paleomagnetism to help to date the time of acquisition of natural remanent magnetization (NRM). Dave Strangway found a clever way of using the idea from one of these tests to help interpret the NRM of the lunar samples, whose orientation in the regolith was known. It is called the conglomerate test. On earth it is used to test the stability of the NRM of a sedimentary unit. One measures the NRM of clasts (bits) from the unit of interest that have been eroded and ended up in a younger sediment. Often that sediment will be a coarse lithified mixture, called a conglomerate. If the NRM of the individual clasts in the younger sediment is randomly oriented, this is good evidence that the process of being incorporated in the younger sediment did not reset the NRM of the blocks. Dave applied this idea to say that because the direction of the NRM of those blocks whose orientation in the regolith was known were random, they could not have acquired their NRM in their present configuration on the lunar surface. It must have survived from some earlier event.

7.2 Laboratory Techniques

Laboratory techniques had evolved for the measurement and interpretation of the NRM of terrestrial rocks that could now be applied to the lunar samples. The standard instrument was the spinner magnetometer, in which the changing magnetic field from the sample, as it is spun generates a voltage in a pick up coil. The magnitude and phase of the voltage induced in the coil can be related to the remanent magnetization of the sample. Fortunately, a new high sensitivity superconducting magnetometer came on line at about the time the lunar samples were made available for study. This made measurement of the lunar samples, which were often weakly magnetized, much easier. A discussion of details of measurement techniques is given in Appendix 4.

Having determined the NRM (Natural Remanent Magnetization) of samples, demagnetization methods were used to try to separate and eliminate secondary natural remanent magnetization, which might contaminate the primary remanent magnetization acquired when the rock was formed on the Moon. This is the detective work in the process, or in modern parlance CSI.

The first method is thermal demagnetization. It takes advantage of the different blocking temperatures of the various magnetic carriers that we learnt about in the last chapter. After initial measurement, the sample is placed in an oven, which is shielded to minimize the magnetic field in it (Fig. 7.2). It is heated to the desired temperature and subsequently allowed to cool to room temperature in the "zero" field of the oven. Then it is remeasured. All of the magnetization unblocked during the thermal cycle will have been demagnetized during heating, but not remagnetized on cooling because the sample cooled in "zero" field. By subtracting the magnetization vector after each thermal demagnetizing cycle from the value before, one gets the direction and intensity of magnetization blocked in the temperature range of that cycle. This procedure is repeated to increasing temperatures until the NRM is completely demagnetized. Often the secondary NRM is blocked at lower temperature than the primary signal, so the secondary noise can be eliminated by using the NRM left after thermal demagnetization.

The principle of the second method of demagnetization is similar to the first except that AC generated magnetic fields are used in the progressive demagnetization steps instead of thermal cycling (Fig. 7.3). The method is made possible by the variation in the magnetic field necessary to demagnetize the NRM of individual particles. The procedure is that the sample is cycled through progressively increasing alternating magnetic fields in a shielded "zero" steady field region and magnetization before and after each cycle measured. There are various ways of ensuring that new magnetization is not picked up as the field is ramped down that

Fig. 7.2 Schonstedt thermal demagnetization system with magnetic shielded furnace (Courtesy of Schonstedt management—many thanks)

Fig. 7.3 Schonstedt alternating field demagnetization system—*left to right*, power controls, sample handler, coil and μ-metal shield

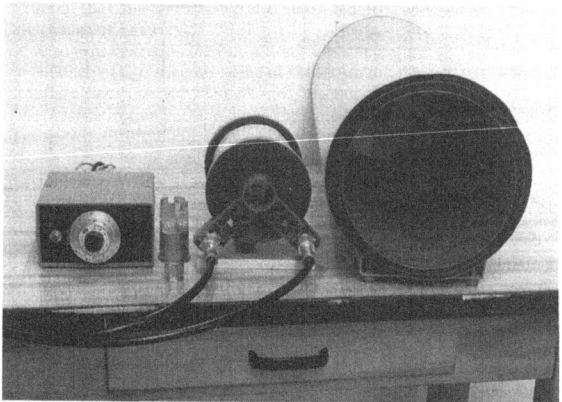

we need not worry about. Again analogously with thermal demagnetization, the secondary noise is often carried by particles that are demagnetized at lower fields than the signal, so it can be eliminated by using the NRM after some AF demagnetization.

7.3 Lunar Paleomagnetism: Beginning to Attack the Puzzle

These were the principal techniques available to us. Now we needed to see whether the NRM of the lunar samples included primary signal that required an ancient dynamo to explain it. Again in the parlance of current TV shows, this was a very cold case. Yet the clues should be there, if we looked hard enough. It might be difficult to establish thermal demagnetization characteristics, but to compare AF demagnetization characteristics of NRM with remanent magnetizations generated in the various possible ways should be straightforward. Then we could eliminate mechanisms of magnetization, whose demagnetization characteristics did not match the NRM.

The first and most obvious way that the NRM might have been acquired was from magnetic fields in the spacecraft. Again Dave Strangway led the way, sending a sample from an earlier mission back to the moon after AF demagnetization, in order to see the possible effects of fields in the spacecraft. The result was very clear. Some new remanent magnetization was picked up on the round trip, but it was very soft and could be demagnetized by weak AF demagnetization fields, so that this was not likely to be mistaken for a primary TRM acquired by the sample when it cooled on the lunar surface.

The next possibility was that while the samples were on earth, they had picked up magnetization, just by being in the geomagnetic field. This is the viscous remanent magnetization (VRM) mentioned in the previous chapter. Given a distribution of grain sizes carrying magnetization, some particles may be unblocked at room temperature or at lower temperature because they are of such a fine grain size that their magnetization is not stably fixed in the sample at room temperature, but is free to follow the ambient field. Others may be close to this size and carry remanent magnetization that will grow in a field over time, or demagnetize over time when the grain is placed in zero field. In multidomain material, wall blocking may be sufficiently weak that they can move over time and change the magnetization of a sample. By the time the samples had been on earth for decades, this magnetization could be significant in some samples. Takesi Nagata and colleagues investigated this problem and were able to divide the samples into two groups on the basis of how susceptible they were to acquiring the VRM. Type I had stable magnetization and they were not very susceptible to VRM. These included most of the crystalline rocks and the melt breccias formed in major impacts. Type II samples, picked up strong VRM that would have been equivalent to the total NRM of the sample after being in the geomagnetic field for decades. Many of the breccias, such as the regolith breccias, formed by shock lithification of the soil, and some of the crystalline rocks with very fine iron fell into this Type II. Fortunately, there was a partial solution to this problem. VRM in these samples could usually be removed by thermal demagnetization to about 100–300 °C, which did not bring about irreversible chemical changes. This became important because in some samples the VRM was very resistant to AF demagnetization. However, it is fair to say that the Type II samples remained problematic and it was best for us to avoid them as much as possible, whereas Type I were farless of a problem, although after 40 years some of these could also be seriously contaminated.

The third possibility was that impact related shock on the moon might account for the lunar paleomagnetic signal. The effect of shock on the magnetic record was one of the first concerns in terrestrial paleomagnetism. It provided ammunition for critics of early studies concerned with testing continental drift. They pointed out that although magnetism was known to be sensitive to shock, samples were frequently hammered out of outcrops. However, this difficulty had been answered by Nagata long ago, who had pointed out the resilience of TRM in typical igneous rocks to shock demagnetization and remagnetization.

7.4 Summary

With the availability of the lunar samples for paleomagnetic studies, several groups began to study their NRM, using the standard methods of the time. It soon became clear that the samples were relatively easily measured, a task made easier by the arrival of a new generation of high sensitivity magnetometers. Now the task was to separate noise from any genuine signal recording ancient lunar magnetic fields. Given that we do not know a priori that lunar rocks have a primary TRM from initial cooling, and given that we do know that essentially all lunar rocks studied were impact-excavated, the possibility of their NRM having been modified, or even dominated by shock effects is a much more open question in lunar paleomagnetism than with terrestrial samples. We know that impacts do indeed generate large shock pressures over considerable volumes of the lunar near surface. Moreover, experimentally generated shock effects observed in terrestrial and lunar rocks have established the characteristics of impact-related magnetization. The question now was to see if there was a signal from lunar primary TRM, which had survived these shock effects. To cover this topic, we will devote the next chapter to impact shock on the lunar surface and its possible effect on the lunar paleomagnetic record.

References

Cournède C, Gattacceca J, Rochette P (2012) Magnetic study of large apollo samples: possible evidence for an ancient centered dipolar field on the moon. Earth Planet Sci Lett 331–332:31–42

Garrick-Bethell I, Weiss BP, Shuster DL, Buz J (2009) Early lunar magnetism. Science 323:365–359

Lawrence K, Johnson C, Tauxe L, Gee J (2008) Lunar paleointensity measurements implications for lunar magnetic evolution. Phys Earth Planet Inter 168:71–87

Chapter 8
Impact Related Shock on the Lunar Surface and the Lunar Paleomagnetic Record

The work of Baldwin and others discussed in Chap. 2 had shown that the craters and basins of the moon were the result of giant impacts, which must have generated large shock waves over considerable volumes. Could these shock effects, or possible magnetic fields generated in the events, explain the remanent magnetism (NRM) of the lunar samples?

8.1 Cratering on the Earth and Moon

With increasing interest in both lunar and terrestrial craters, major progress was made in the Apollo days in understanding crater morphology. Models of their generation were developed and the shock effects on minerals were established as a means of calibrating shock levels. This was important information for us because now we had an idea of the levels of shock to be expected in the lunar samples. The key question was whether impact related shock could account for lunar magnetism, either by magnetizing material in ambient fields, or by generating fields in which magnetism could be acquired.

Crater morphology is illustrated in Fig. 8.1,[1] which shows the twofold division of craters into simple and complex. The distinction is that in complex craters the event has been large enough to generate a central peak, which is a rebound effect similar in origin to the rebound seen in the familiar high speed photos of impacts of milk droplets into milk. In the complex crater the central uplift consists of shocked target rocks.

In addition, there are giant basin forming impacts giving rise to multi-ring basins on the scale of a thousand kms across (Fig. 8.2).[2] One can't help wondering

---

[1] Courtesy NASA image but see also Horz et al. (1991). In addition to the detailed discussion of crater formation, there is an important summary of shock metamorphism in this same chapter. The Lunar Source book itself is an invaluable resource for lunar scientists and has been followed by Jolliff et al. (2006), that provides an updated volume of similar caliber.

[2] Courtesy NASA image lunar orbiter.

Fig. 8.1 Simple and complex crater morphologies

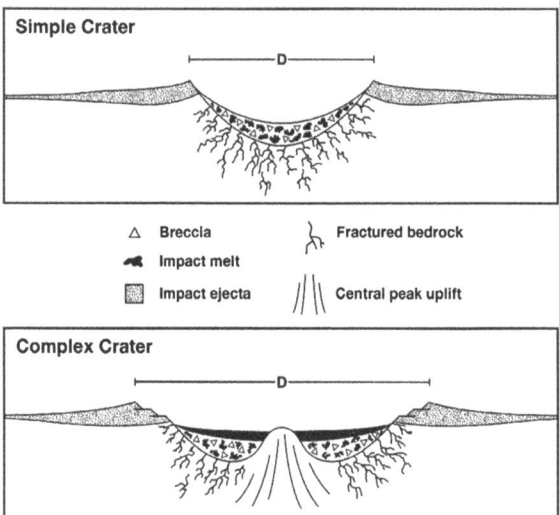

Fig. 8.2 Mare Orientale—a giant multi-ring basin

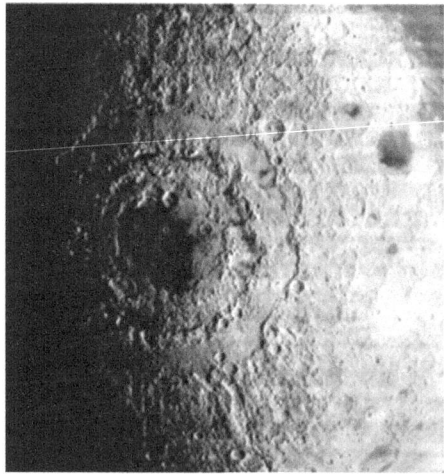

at the mythology that would have arisen to explain this basin had it faced the earth, rather than being as it is on the limb. The increased understanding of the impact process on the large scale made it possible to start thinking about the scale of the shock to be expected from giant impacts.

The Fig. 8.3[3] shows schematically the effects of the impact of a km sized projectile travelling at 10 km/s. In the initial stage of contact and compression, we see enormous shock levels that produce vaporization and melt close into the

---

[3] Courtesy Lunar and Planetary Science Institute. See also French (1998).

8.1 Cratering on the Earth and Moon

Fig. 8.3 Initial shock wave pressures and effects of shock on rocks

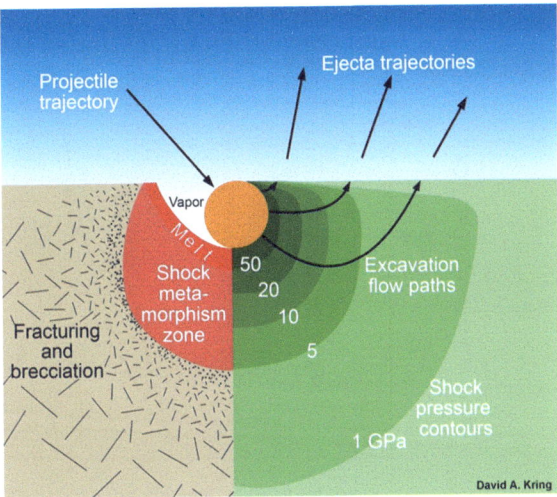

impact point and further away to tens or hundreds of kms massive disruption of the crustal rock. In the figure you see shock pressure contours and the unit of GPa (Giga Pascal). Units can be a nightmare in science with different systems evolving over time and even though a universal system was proposed in the last century. Appendix 2 gave the metric prefixes and a few examples of units we will use. To avoid the confusion of units as much as possible, we will again use comparisons. It turns out that 1 GigaPascal is equal to ∼10,000 atmospheres, and a few GPa is equal to the pressure at a depth of about 100 Kms down in the mantle. With this background, the key task now is to understand what these shock pressures do to lunar rocks and their magnetization.

The shaded area in the bottom left of Fig. 8.4,[4] bounded by pressures of up to a few GPa and temperatures of 1,000 °C, is the province of the geologists, as they study the effects on rocks that have been buried in the earth and their constituent minerals metamorphosed to high temperature and pressure forms. The area to the right, which reaches much higher temperatures and pressures, is the province of shock metamorphism, where we encounter the products of hypervelocity impacts, such as those on the moon and in terrestrial impact craters. The near vertical lines show the boundaries between the ranges of graphite and diamond, quartz and coesite and coesite and stishovite. With increasing pressure, the low pressure forms invert to the high pressure forms. You may remember that the occurrence of coesite at Meteor crater and at other impact sites demonstrated to Chao, Shoemaker and their colleagues that these were indisputably impact craters. One does not see shock levels of GPa in volcanic events, but shatter cones are also found at impact sites such as Vredefort implying shocks of a few to 10 s of GPa. With the laboratory calibration shown in Fig. 8.4 we can find the temperature and pressures

---

[4] Courtesy NASA Goddard Space flight Center Tutorial Image. See also French (1998).

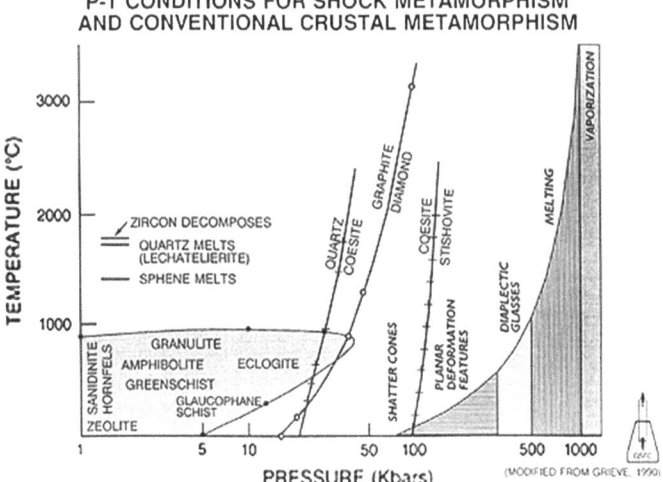

Fig. 8.4 A plot of temperature against pressure for normal geological metamorphism and shock metamorphism

involved in shock metamorphism of lunar rocks and hence the pressures and temperatures involved in lunar impacts. This allows us to estimate the effects of shock associated with impacts on the lunar paleomagnetic record, providing we know the effect on magnetization that particular levels will bring about.

8.2 Shock Effects on Remanent Magnetization

As a shock wave passes through a solid, the first phase is the passage of an elastic deformation region. At low shock levels of few GPa no plastic deformation will occur, as the shock is insufficient to exceed the Hugoniot—the boundary between elastic and plastic deformation. At higher shock levels of roughly >10 GPa, the elastic deformation region, will be followed by the main shock, where the Hugoniot level is exceeded. Petrographic indicators of the plastic deformation have a variety of forms such as fractures, twinning, and planar deformation features disrupting crystals then occur. As tens of GPa are reached glass is formed and by 70–80 GPa melting is pervasive. These values relate primarily to solid rocks and by contrast in powders melting sets in at much lower temperatures (see footnote 1). Clearly at the levels of shock in which plastic deformation has occurred the magnetic microstructure controlled by dislocations and other deformation features will be affected. Finally, temperature and in particular the residual temperature after the wave has passed will become dominant.

First, we look at the classical interpretation of shock remanent magnetization at low pressures that follows from the application of static stress. The application of

8.2 Shock Effects on Remanent Magnetization

stress in the presence of a magnetic field tends to increase the magnetization in the direction of the field. This is because in addition to all of the other energy terms giving adjustments to the magnetization in response to the stress, the magnetic field term is minimized when the magnetization is parallel to the field. This magnetization was termed piezoremanent magnetization (PRM). In the absence of an ambient field, the result is demagnetization. Here the rearrangement of the magnetization favours the reduction of magnetization because this reduces the magnetic energy of the particle. Multidomain magnetic particles are in general much more sensitive to stress than single domain particles. PRM at low stress levels ($<\sim 0.1$ GPa) will tend to be predominantly soft and in multidomain grains. The best way to demagnetize a PRM is then via AF demagnetization.

Shock remanent magnetization (SRM) in the same low stress range behaves similarly. At greater shock levels, the samples exceed their elastic limit and suffer permanent deformation, as the shock wave passes through them. This is accompanied by the possibility of petrological markers of shock levels, as we saw in Fig. 8.4. In this range of a few tens of GPa, there are complicated effects of shock on remanent magnetization, so we avoid such rocks in seeking reliable paleomagnetic recorders. With still greater shock the rock can be melted and gain a TRM when it subsequently cools. Hence these rocks are again potentially good recorders for us.

There was one special case that intrigued Stan Cisowski and I, as we started work on shock effects on the lunar samples and that was the formation of the regolith breccias by impacts. We knew that they were the results of shock compaction of the lunar soil, so why not try to make some artificial regolith breccias by shocking lunar soil. Then by comparing the magnetic properties of our artificial regolith breccias with the properties of the real breccias we should get a rough idea of the shock levels that formed them. We could also see if they acquired magnetic remanence in the process and hence get another record of ancient lunar fields.

Stan and I joined with Pete Wasilewski of NASA Goddard to get in touch with Dr. M.E.Rose at the Naval Weapons Laboratory at Dahlgren in Virginia and told him what we had in mind. He immediately said that the flying plate technique that they used should work perfectly for the shock lithification of the lunar soils. Figure 8.5 illustrates the sample holder and the sample assembly for flying plate experiment. Our lunar soil samples were packed in the holder and placed in the assembly as illustrated. A Du Pont sheet explosive line wave generator is placed between two glass sheets and canted. When it detonates, the glass shatters and moves downward as a plane reaching the surface of the main explosive charge simultaneously. The main charge then explodes and propels the driver plate into sample holder. The shock wave then travels through the sample holder as a near plane wave. Four samples were used in each experiment. The samples were stopped in sawdust traps and their orientation recorded. We therefore knew the orientation in which they were shocked and in which they were recovered.

The properties of the synthetic breccias covered the range of the petrological and magnetic properties of the Apollo regolith breccias. The magnetic properties of the poorly consolidated regolith breccias like the Apollo 14 sample 14047 were similar to those of soil shocked to <1 GPa, whereas the well indurated breccias

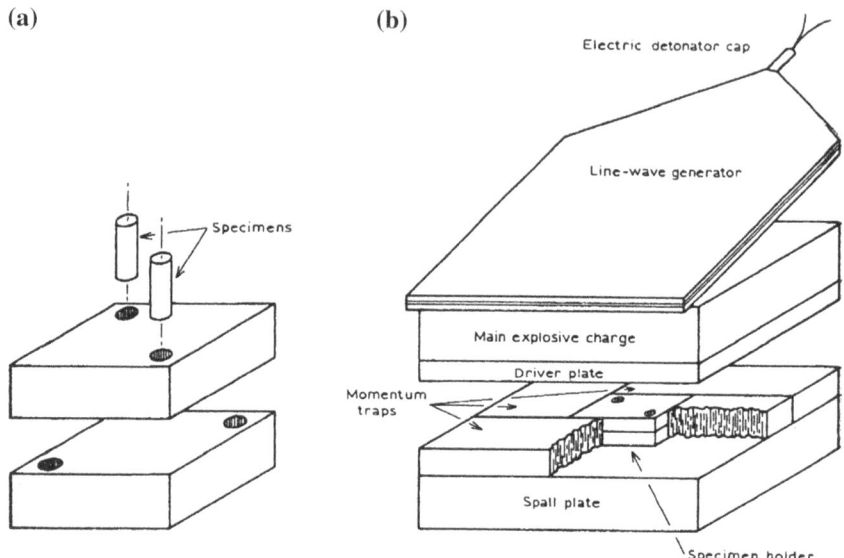

Fig. 8.5 a Left sample holder and right sample holder assembly. **b** Direction of SRM for 7.5 GPa shocked soil on AF demagnetization (Cisowski et al. 1973)

like the Apollo 11 sample 10048 and the Apollo 15 sample 15498 were consistent with shocks of about 5 GPa. It appeared that soil compression in the formation of the breccias took place in shocks of a few GPa. The match we found in magnetic properties was consistent with other ideas of the shock levels expected in regolith breccias. Although some of our colleagues have rightly pointed out that we did not start with well controlled material and we don't know accurately the shock in the actual sediments as opposed to the metal sample holders, the results do give an indication of the shock involved. The directions of the remanent magnetization tended to be intermediate between the field during shock and during cooling after the event. Additional experiments carried out at the NASA Ames vertical gun ballistic range showed that the synthetic breccias formed there were similar to those made at the Naval Weapons laboratory and acquired magnetization roughly in the direction of the field in the chamber, but that they did not record the intensity accurately in this shock range.

While the possibility of getting a reliable record of fields out of the regolith breccias in general was not promising, other breccias were very promising. These were the melt rocks and melt breccias. In their formation, they had been heated well above their Curie points and so at least initially they must have carried a TRM as they cooled, just as the mare basalts did when they initially cooled. The other good news was that most mare basalts showed no petrologic signs of shock. This was not to say that they had not been shocked; our petrologist friends could only detect shock effects at a level of a few GPa, which we knew was strong enough to

8.2 Shock Effects on Remanent Magnetization

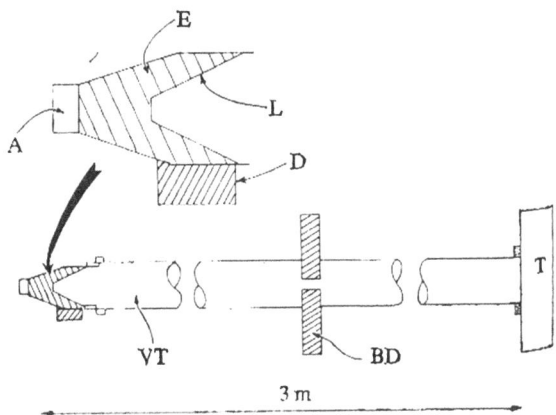

Fig. 8.6 Experimental set up for shaped charge acceleration of tip into the target rock. *A* detonater, *E* explosive shaped charge (octol), *D* deflector (TNT). *L* aluminium liner, *BD* blast deflector, *T* downrange rock target (Cisowski et al. 1973)

affect magnetization. The next step was obviously to see the effects shock on solid rocks, such as basalts, melt rock and melt breccias.

Solid rocks are not as susceptible to effects of shock as the soil, in which individual particles move relative to each other, absorb more energy, and generate more heat than do solid rocks. Earlier shock work on rocks had tended to be in lower levels of shock than would have been experienced on the moon, so again we got in touch with an established group with experience in high shock level work. This was Giuseppi Martelli's group from the University of Sussex, which operated a facility outside Rome. The experimental set up is illustrated in Fig. 8.6.

Figure 8.7a shows one of the large craters formed in the target Rowley Regis basalt. Samples were taken at various distances from the center of the crater to see the pattern of acquisition of remanence, as a function of the distance to the centre of the crater. Just as in the breccia experiments, the magnetic properties of the basalt were changed by the impact depending upon the shock experienced, which in this case depended upon the distance from the center of the impact crater. The control sample appears to carry stable, NRM which is TRM like with some low field noise (Fig. 8.7b). By demagnetization to 70 mT, its NRM is far more stable than IRMs magnetization as we would expect in a fine grained basalt like this.

The sample from near the center of the crater has completely different AF demagnetization characteristics (Fig. 8.7c). Moreover, its saturation remanent magnetization had increased showing that the shock has changed the rock significantly. The ratio of SRM to IRMs is more or less constant apart form the initial measurement. It is clear that shocks in the range of around 10 GPa, as in this experiment, will generate strong SRM giving identifiable changes of demagnetization characteristics in solid rock and modify the rock magnetically. This sample is however, fine grained whereas most lunar rocks we are deal with are coarse grained and multidomain, so changes will take place but they may not be the same. We will have to be careful with our fingerprints.

Another aspect of these shaped charge experiments was intriguing. Magnetic field observations were made to see if the ionized particle cloud or plasma

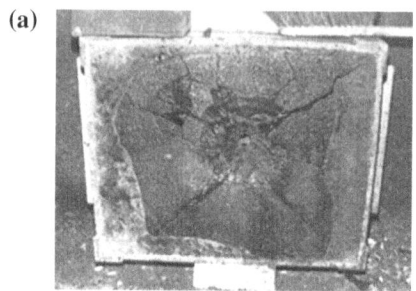

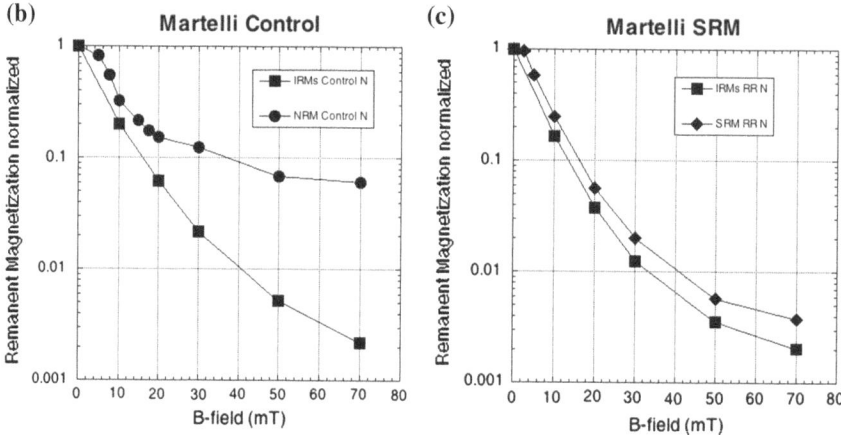

Fig. 8.7 Rowley regis basalt (**a**) Crater formed by the impact ~ 18 km/s. Block size ~ 50 cm. (**b**) Control sample showing normalized *NRM* and saturation remanent magnetization (*IRM*s) and (**c**) Normalized shock remanent magnetization (*SRM*) and saturation remanent magnetization at a distance from crater center of 4 cm (Srnka et al. 1979)

generated by the impact had any effect on the local ambient field. These showed that in the millisecond after the impact, the applied field was amplified by nearly a factor of three. This result bears upon the second question raised at the beginning of this chapter—could magnetic fields be generated by impacts, and if so, could they be recorded by the target rocks. The most likely explanation of the field change is that the expanding plasma of charged particles formed by the impact explosion had compressed the applied field as illustrated in Fig. 8.8. A similar effect is seen during the explosion of nuclear weapons in the atmosphere, when the expanding fireball, or plasma cloud, excludes the geomagnetic field and compresses it. Since only one component of the magnetic field was observed and simple compression appears to account for the observation in this experiment no other impact generated field is required to explain the result.

Experiments by Crawford and Schultz (1991) of impacts of small iron projectiles into powdered dolomite at by velocities of 5.75 km/s gave evidence of impact generated fields. These fields varied on the time scale of microseconds.

8.2 Shock Effects on Remanent Magnetization

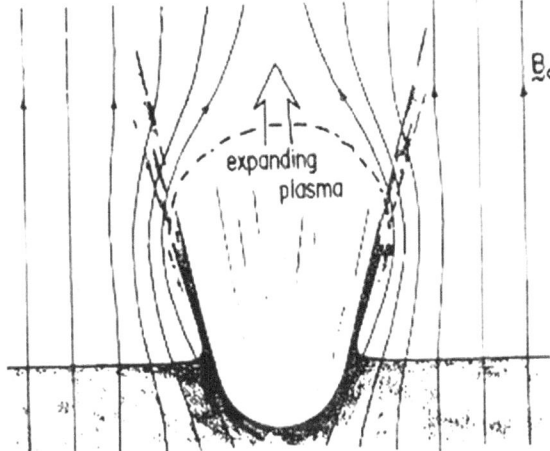

Fig. 8.8 Model of the effect of the impact explosion on the ambient field—B_0

However, the fields were weak. Some theoretical studies suggest much stronger fields may be generated. However, that impact generated magnetic fields play an important role in giving the lunar paleomagnetic record is unproven.

In addition to the laboratory experiments, there was some evidence from the early observations of the locations of the major magnetic crustal anomalies on the moon that they might be related with impact generated fields. This evidence came from the antipodal location of several of the strongest magnetic anomalies to Orientale, Imbrium, Serenitatis and Crisium basins found by the electron reflectance experiment as described towards the end of Chap. 5. A model was developed whereby the expulsion of field at the impact site similar to that shown in Fig. 8.8 would sweep an ambient field on the moon across the surface of the moon until it coalesced at the antipode (Hood and Huang 1991). Mechanisms of magnetization at the antipode must take no longer than the duration of the field, so TRM of a large volume is not possible. However, remanent magnetization from pressure (PRM) or shock (SRM) effects in the seismic wave crossing the moon, or a mechanism involving the deposition of ejecta are both possible. The idea has been made less attractive recently by the reanalysis of the South Pole Aitken basin anomalies, which we will meet in our final chapter and will show that some of the features interpreted as antipodal sources were in fact related to local features in South Pole Aitken.

Among the novel approaches in recent studies of SRM has been the use of lasers to generate shock on a microscopic scale, so that the effects can be studied with optical, electron and magnetic microscopy.[5] Shock waves reaching pressures of hundreds of GPa can be produced in regions a few mm in diameter with high powered pulsed lasers. The shock wave penetrates into the sample followed by the

---

[5] Hood and Huang (1991), Hood and Artemieva (2008). Another important contribution was the analysis of central peak magnetic anomalies in older impact basins providing evidence for an early dynamo—Hood (2011).

rarefaction wave nanoseconds later. The results of experiments with a magnetite bearing basalt yielded a low over the crater and overall anomalies that look very similar to the mapped magnetic fields over lunar craters. Thus despite the enormous scale difference between the experiment and nature similar results were obtained.

It has also been confirmed that SRM is less efficient than TRM and that it is predominantly soft and easily demagnetized by the AF method. The experimental work established that the efficiency of acquisition of SRM was about 10 % of TRM in the range of up to ~ 1 GPa. When a sample carrying TRM was subsequently shocked in a magnetic field to the same range of ~ 1 GPa, its magnetization increased substantially. However, with AF demagnetization the SRM disappeared and the TRM remained.[6] These modern experiments clarified the effects in lunar samples in the range of interest and it became clear that in the range of shock approaching the level at which petrologists could detect shock effects the SRM was soft and easily demagnetized by the AF technique.

8.3 Magnetic Fields of Terrestrial Impact Craters

The magnetic fields of many impact craters on earth have been studied. The magnetic signature of small ($<\sim 40$ km) terrestrial craters tends to be a magnetic low, but in larger craters there is often a central high (Fig. 8.9). These central peak anomalies have been variously attributed to TRM carried by melted and heated rocks, CRM due to impact-induced hydrothermal alteration of the fractured, heated target rocks, and even, occasionally, to SRM. Impact-related TRM acquired in the Earth's field has been clearly identified in many crater melt rocks and shown to be consistent with the ambient geomagnetic field. However, there is as yet no completely convincing evidence for the presence of SRM. Similarly, there is no clear evidence for NRM acquired in impact-generated fields, although the constant presence of the geomagnetic field makes paleomagnetic studies highly insensitive to any such weak fields. Nevertheless, we can be confident that no very strong fields have been recorded in terrestrial craters.

8.4 Summary

From the studies of shock remanent magnetization and of cratering effects, it became clear that we needed to concentrate on the minimally shocked samples, as indicated by the petrological studies, and samples that had been very strongly shocked so as to melt and recrystallize to give virtually new rocks. We shall see that Mare basalts, the melt rocks and melt breccias were the most promising material for a paleomagnetic

---

[6] Courtesy of Ernstson Claudin and Fernando Cladin with many thanks.

Chapter 9
Lunar Paleomagnetism and the Case for an Early Lunar Dynamo

Criteria for reliability of the lunar NRM are established and the test of the case for an early dynamo described. The history of the intensity of ancient lunar fields from the samples is the key observation we need and unfortunately even more difficult than determining paleomagnetic direction reliably, but progress has been made. The results from the samples are tested against the record from the lunar crustal magnetic anomalies. Evidence for a lunar dynamo grows. In this long chapter, we will have to go further into the nuts and bolts of paleomagnetism and the interpretation of magnetic anomalies than before in order to test the case for an early dynamo in the lunar core. I hope that the background built up so far will stand the reader in good stead. Fasten safety belts!

9.1 Criteria for Reliable Recording of Ancient Lunar Fields by NRM

To test the case for an early lunar dynamo, we first need to set up criteria to be met by samples before we accept them as carrying a legitimate record of lunar fields. The first and key criterion follows from the last chapter. There should be no petrologic evidence of shock of >5 GPa. Admittedly this is an arbitrary cut off, but at least it is possible for our petrological colleagues to recognize this level of shock by its effect on the constituent minerals in the lunar samples. Moreover, we know that even if samples have experienced such shock, the Shock Remanent Magnetization (SRM) they carry will be soft and easily removed by AF demagnetization. Second AF, or thermal demagnetization should be successful in isolating a single direction of NRM (Natural Remanent Magnetization), as demagnetization continues to completion. The third criterion requires agreement in direction and intensity of NRM between sub-samples from a single rock sample. Fourth and finally the AF demagnetization, or thermal demagnetization characteristics of NRM should be consistent with Thermal Remanent Magnetization (TRM) and can be distinguished from at least, Shock Remanent Magnetization (SRM), Viscous Remanent Magnetization (VRM), or weak field Isothermal Remanent Magnetization (IRM) contamination, i.e. the NRM should have the demagnetization fingerprint of a TRM that tells us that it carries a bona fide record of the field when it acquired this NRM.[1]

The first three criteria follow from previous discussions, although the directional analysis has been shortchanged, because presenting plots of 3D directions on a 2D piece of paper opens another can of worms, and so we won't illustrate them, we will simply describe the results in words. The fourth criterion needs some more explanation, although we have met the idea before. Several times we have encountered different behavior on AF demagnetization associated with different types of the remanent magnetization picked up from fields experienced in the spacecraft by the sample returned to the moon by Dave Strangway was very soft. It was easily demagnetized. In contrast, we have seen examples of remanent magnetizations that are very stable against AF demagnetization, e.g. TRM (Thermal Remanent Magnetization). For a number of samples, we now have these AF demagnetization characteristics for different remanent magnetizations, so that we can make comparisons using the behavior as fingerprints for particular mechanisms of remanent magnetizations.

Let us now see whether any of the samples analyzed pass the criteria we have set up and justify the interpretation of a primary NRM, acquired as they cooled initially on the moon. Fortunately, several of the samples that gave promising results earlier have now been reanalyzed using modern techniques. This work has been primarily by Ben Weiss's MIT group and has yielded far more convincing data. The principal improvements were in the AF demagnetization techniques. First, much higher demagnetizing fields became available, which meant that the signal carried by fine Fe particles with high coercivity magnetization could be analyzed. These were the best hope for high fidelity records. Second, automatic sample handling mechanisms were developed that permitted magnetometers to be run for long periods unattended. This permitted much more detailed observations than were feasible previously, when loading and unloading samples after each measurement was done manually. Third more sophisticated analysis methods and software were developed for AF demagnetization. The principal developments in all these areas were pioneered in the Caltech paleomagnetism laboratory under the leadership of Joe Kirschvink, who was also Ben Weiss's PhD advisor.

9.2 Apollo 11 Mare Basalt 10020

We will look at just one sample in detail as an example of the new work. First note that the original results from 10020 in the Apollo days suggested that it came near to satisfying the criteria for acceptance given above. It is one of about half a dozen

---

[1] These criteria were set up with Ben Weiss and are discussed in a review paper *Impact related shock remanent magnetization and lunar paleomagnetism*. In Preparation.

9.2 Apollo 11 Mare Basalt 10020

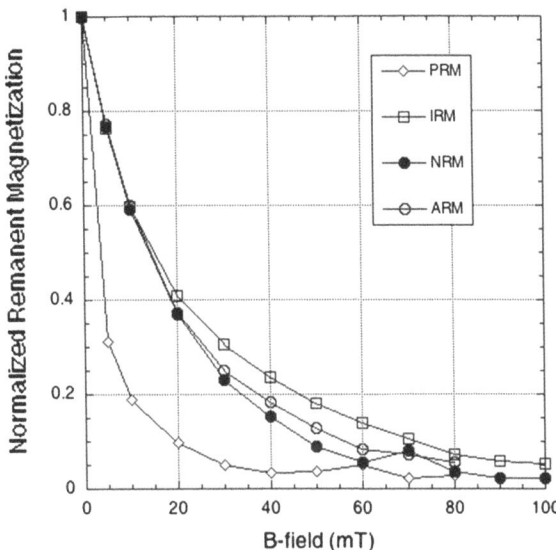

Fig. 9.1 Comparison of AF demagnetization fingerprints of various remanent magnetizations in mare basalt 10020 redrawn from data of Shea et al. (2012)

samples, which have now been reexamined. In a recent study,[2] multiple samples were investigated from 10020.234 and AF demagnetization was much more detailed and taken to higher fields than in the earlier work. The directions and strength of hard and soft components isolated in AF demagnetization could then be compared. The soft components in different samples are in very different directions, but the directions of the hard components are statistically indistinguishable. This is similar to the experience with other lunar samples and suggests that the sample picked up secondary NRM in different directions probably during sample preparation, but with demagnetization this was eliminated leaving a consistent high coercivity component apparently recording the ancient lunar field at 3.7 Ga.

The AF demagnetization characteristics are shown in Fig. 9.1. Two of the magnetizations compared are NRM and IRMs, which we have already met. In addition results are shown for PRM, which is the static stress or Piezo (pressure) Remanent Magnetization, mentioned in the last chapter. Remember that in the low shock range the behavior of SRM and PRM are similar. The NRM fingerprint clearly differs from this.

ARM stands for Anhysteretic Remanent Magnetization, which is a good, though not perfect, proxy for TRM. It has the great virtue that it is given without heating. To give ARM the sample is treated as if it were to be AF demagnetized, but when the AF field is finally reduced, it is in the presence of a small magnetic bias field. Notice that NRM and ARM behave similarly. Both are a little softer than IRMs, consistent with the signal coming predominantly from multidomain

---

[2] Shea et al. (2012). This paper presents the evidence for the continued operation of the lunar dynamo until 3.7 Ga.

grains. Although it is not indicated in the figure, these results are based on much more detailed AF demagnetizations than would have been feasible in the Apollo days. The directional analysis was also correspondingly more convincing.

The new results make a good case for a primary NRM recording the field, when this lava cooled on the lunar surface. There is no evidence of strong shock. AF demagnetization isolates a single high coercivity direction similar in different subsamples. The demagnetization fingerprint of the NRM is similar to the TRM proxy and dissimilar from the weak field IRM or weak shock. It therefore passes all the criteria outlined above.

9.3 Intensity of Recorded Fields from NRM

Having satisfied ourselves that 10020 is a valid recording of the lunar field at 3.7 Ga, we now need to find the intensity of the field recorded. As we noted at the outset, the intensity of paleomagnetic fields has always been more difficult than determining their direction. All these methods of intensity determination are based on replacing the NRM of the sample with another remanent magnetization, such as TRM, in a known field.

Then one can uses the following simple ratio to find the intensity

$$Ancient field / Laboratory field = NRM / Laboratory TRM.$$

We know the laboratory field, the intensity of NRM and of the laboratory TRM, leaving as the one unknown the ancient field. The problem is to give TRM in the laboratory field to the lunar samples without bringing about irreversible changes that destroy the validity of the experiment.

This problem was so difficult that Stan Cisowki and I decided it would be useful to get a very rough idea of the history first by any possible means rather than wait for the success of classical methods. We had been using the ratio of NRM: IRMs at 20 mT demagnetization to help understand NRM in different samples using the fingerprint method described above. By this time we had plenty of samples analyzed in this way. He suggested that we should calibrate the ratio as a means of estimating even roughly the intensity of the fields recorded by the relatively large number of samples for which these data were available. He determined the calibration factor from results of the relatively few lunar samples for which classical intensity determination were available (Fig. 9.2).[3]

---

[3] Modified from Cisowski, S. M., and Fuller, M., 1986, *Lunar paleointensities via the IRMs normalization method and the early magnetic history of the moon*, pp 411–424, In Hartmann, W. K., Phillips, R. J., and Taylor G. J, Eds. 1986. *Origin of the Moon* and Cisowski, S. M., Collinson, D. W., Runcorn, S. K., Stephenson, A., and Fuller, M., 1983, A review of lunar paleointensity data and implications for the origin of lunar magnetism, Proc. 13th Lunar and Planetary Science Conference, J. Geophys. Res., 88, S, A691- A704.

9.3 Intensity of Recorded Fields from NRM

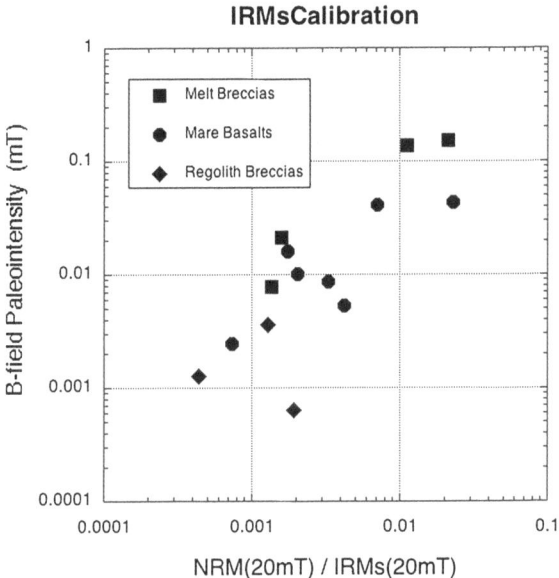

Fig. 9.2 Calibration of IRMs normalization method with lunar samples: Mare basalts 10049, 10050, 10069, 12022, 15495, 15535, 70017; Regolith breccias 15498, 60255, 70019, Melt breccias and melt rocks 62235, 62295, 67915, 68416

The calibration factor to give the field in microTesla (μT) from the ratio of NRM/IRMs at 20 mT was 4700. Recognizing the problems of giving TRM to lunar samples, we needed to test this calibration. To do this, 41 terrestrial igneous rocks were given TRM in a field of 0.02 mT (20 μT) and their TRM and IRMs after 20 mT AF demagnetization measured. The ratio of TRM (0.02 mT) (20 mT)/ IRMs (20 mT) for 80 % of the samples gave an answer within a factor of five of the field used. While this is not the type of accuracy to make anybody very happy, it did justify using the method as a first cut at the problem of the intensity of the ancient lunar field. The highest fields recorded by these lunar samples around 3.7–3.9 Ga according to the normalization method are stronger than the present geomagnetic field, which at ~ 50 μT, yields a ratio of ~ 0.01 for NRM (20)/IRMs (20). Whether there was only a field at this time and everything else was noise, or whether the field grew to this value prior to 3.9 Ga and decreased later was not initially clear. An obvious weakness of the approach, politely pointed out by our colleagues and recognized by ourselves, is that it assumes all NRM at 20 mT demagnetization was primary TRM and that all secondary magnetization had been eliminated. This was clearly not likely to be true, but the result set the stage for classical paleointensity studies (Fig. 9.3).

In the Apollo days, classical paleointensity methods of the time were used and yielded some results, but none passed modern standards with the possible exception of the melt breccia 62235. The results are given in Fig. 9.4, and a somewhat similar pattern of interpreted field strength is seen as came from the IRMs used normalization.

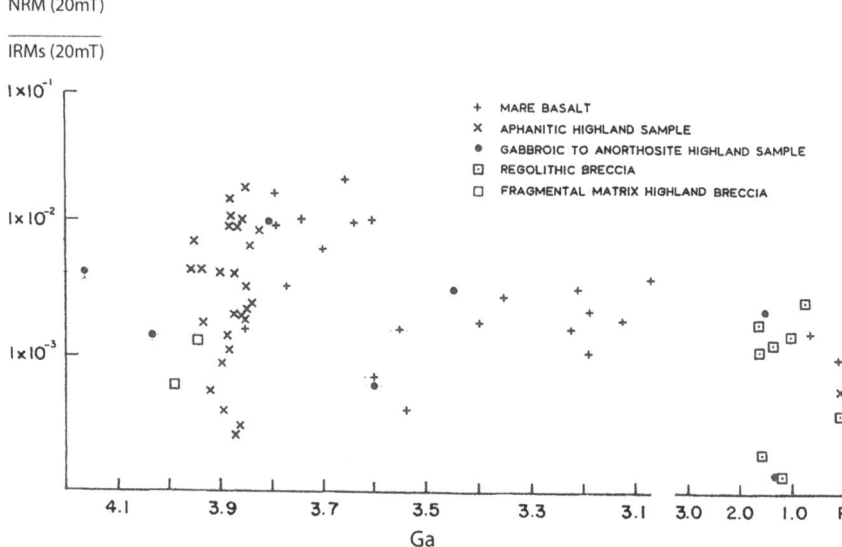

Fig. 9.3 Paleointensity estimates by the REM, or IRMs normalization technique. B-field (μT) ~ 4700 × NRM (20mT)/IRMs (20mT). Ga = billion years

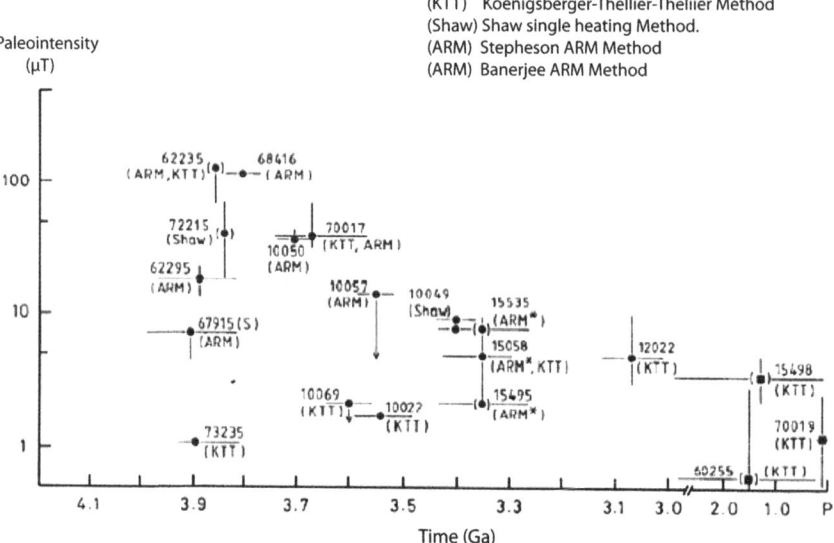

Fig. 9.4 Standard paleointensity determinations from the Apollo era

9.3 Intensity of Recorded Fields from NRM

However, there are many obvious inconsistencies. The various methods are now described and recent applications of them presented.

All paleointensity intensity methods are based upon replacing the NRM with a laboratory remanent magnetism and using the ratio of the measured intensities to get the ratios of the laboratory to ancient fields. The simplest experiment is to demagnetize the sample thermally to completion and replace the NRM with a laboratory TRM in a known field. Attempts to do this in the 1930s, led Thellier in France and Koenigsberger in Germany to a method of stepwise replacement NRM by TRM. In a common manifestation of this method, one cycles a sample in zero field through a small temperature range, and measures before and after cooling to determine the amount of NRM blocked in this temperature range. The sample is then taken through the same temperature cycle and allowed to cool in the known laboratory field to obtain the pTRM in this same temperature range. If we know the laboratory field, the ratio of the NRM to the laboratory TRM gives the ancient field, as we saw above. It became the standard method of paleointensity determination and would have been used exclusively on the lunar samples were there not the problem with heating lunar samples. Fortunately the method did seem to work with some samples and most spectacularly with the impact melt breccia 62235.

Numerous workers had shown that 62235 appeared to carry at least in part a primary NRM, which was thermal in origin. The similarity of the paleointensity results from these samples studied over 30 years in different laboratories is remarkable. When the standard Koenigsberger-Thellier-Thellier method of paleointensity is used, the NRM and TRM in individual temperature cycles yield multiple ratios of NRM/TRM. The results are usually shown in plots of the NRM remaining after the thermal demagnetization in each cycle against the pTRM gained by the end of the cycle. If NRM is a TRM, a straight line should fit the data and its slope will give the ratio of the fields as in Fig. 9.5.[4]

The highest quality data (Fig. 9.5) came from Lawrence and others from UCSD, and were included in the paper questioning the existence of a lunar dynamo. Their data suggest that a pTRM from a temperature of $\sim 500\ ^\circ C$ took place in a field of 92.9 μT. The age of this rock is 3.9 Ga, so that this strong field value falls into the age interval of the strongest fields by the IRMs normalization results. There is also a much weaker field indicated by the change to a lower slope at around 500 °C. This yields a value 2.9 μT for weaker field magnetization blocked above this temperature.

A single heating to acquire the laboratory TRM is fine as long as it does not produce the infamous irreversible changes in magnetic properties. To take advantage of this John Shaw from Liverpool University developed a method with

---

[4] I have always liked to refer to this method as the Koenigsberger-Thellier-Thellier method of paleointensity determination because Koenigsberger appeared to have recognized the essence of the method. Indeed Nagata told me that Koenigsberger had sent him laboratory notes covering this work as he feared that because he was Jewish he would not survive World War II. The method was used on lunar sample 62235 by Sugiura and Strangway (1983), Collinson et al. (1973), Lawrence et al. (2008).

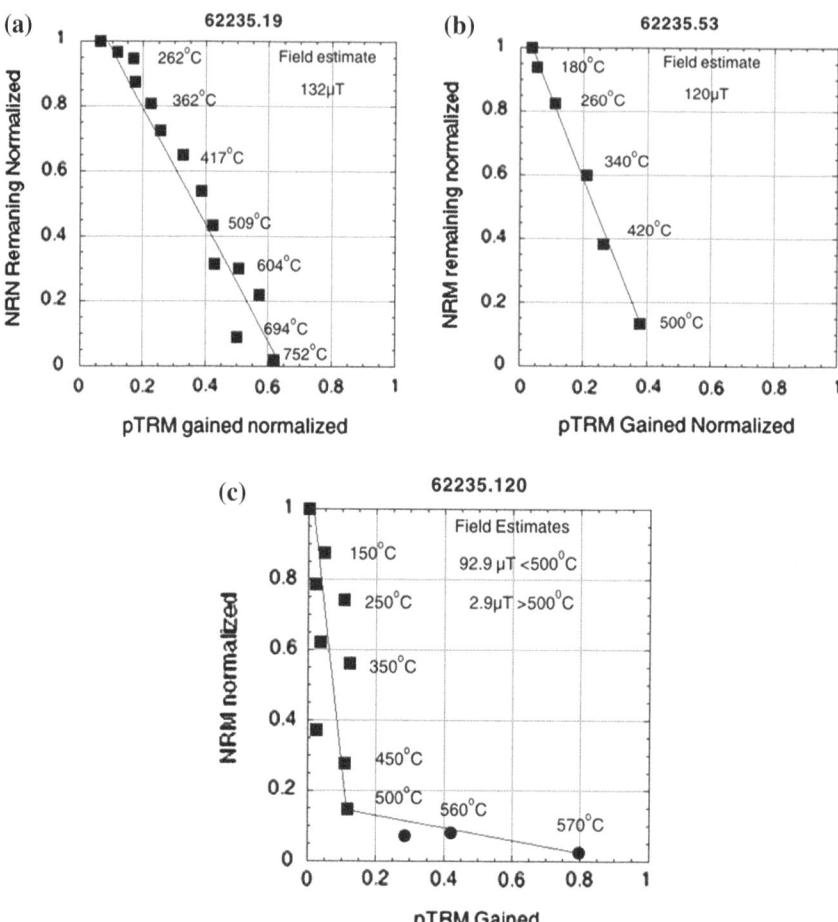

Fig. 9.5 Paleointensity results for 62235 redrawn from data of **a** Sugiura and Strangway (1983), **b** Collinson et al. (1973) and **c** Lawrence et al. (2008)

sufficient tests for these changes that permitted paleointensities to be obtained successfully from some samples. The method is now known as the Shaw method and was used in some of the early work (Fig. 9.4).

Returning to the recent studies of 10020 by the MIT group (2), the use of methods with replacement of NRM by the TRM proxy of ARM (Anhysteretic Remanent Magnetization) yielded important results. The procedure is analogous to the classical KTT paleointensity method, although it is the comparison of the ARM acquired and NRM demagnetized in different ranges of AF demagnetization that are compared. The problem with this method is the need to for calibration because we are not comparing NRM identified as TRM in origin with laboratory induced TRM, but with ARM. We therefore need to know the equivalence

9.3 Intensity of Recorded Fields from NRM

between ARM and TRM. The ARM method was however calibrated, as was the IRMs normalization.

Combining all the results from the modern ARM and IRMs experiments, a paleointensity of 63 µT was obtained for 10020.234. This is similar to the result of 86 µT obtained by the IRMs normalization method in the Apollo days. Admittedly there are major uncertainties in the calibration factors of the ARM and IRMs methods, but the results from the earlier work and the much more comprehensive recent studies make a very strong case for a field of some tens of µT on the lunar surface, when this basalt initially cooled some 3.7 Ga years ago. This is similar to the present surface field on earth.

Comparable studies have been carried out recently with other samples. The result for 76535 by Garrick-Bethell[5] and colleagues is particularly important. With an age of ~ 4.2 Ga, it may turn out to be the oldest unshocked lunar sample available to us. An important aspect of the work on this sample was the detailed analysis of its thermal history. It was shown that the Argon system, which is used for the age determination was closed in the rock soon after cooling and that subsequently the sample has not been heated significantly.

Paleointensities were obtained by Clement Suavet and colleagues (Suavet et al. 2012) for 10017 and 10049 using the same methods as with 10020. 234 (2), and yielded similar values of 10's of µT. If these results are confirmed with additional samples, then the high field values must have started within a few hundred million years of the formation of the moon and lasted for almost a billion years. There is one piece of possibly confirmatory evidence for the early start of the dynamo. In Fig. 9.4, the sample 78155 gives a similar result. However, it has been shocked and has not yet been given the detailed analysis of its thermal history similar to 76535.

With sample 12022 whose age is 3.2 Ga, we see something very different. An earlier study had suggested that this basalt might carry a reliable record of the field at 3.2 Ga, but again the earlier demagnetization was not carried to high enough AF demagnetizing fields. The new result by Tikoo and others[6] showed that the sample's most stable NRM appears to have been acquired in a very weak field, which had not been reported before. If this value is correct, the lunar field has decreased by about a factor of 10 by 3.2 Ga and this field could come from local surface fields rather than from a lunar dynamo. Clearly such an important suggestion needs to be tested against other younger rocks.

There is some additional evidence supporting the interpretation of a declining dynamo between the time of the Apollo 11 and 12 basalts. It comes again from our use of AF demagnetization characteristics of the NRM as a fingerprint for the mechanism of magnetization of samples. Figure 9.6a gives AF demagnetization of

---

[5] Garrick-Bethell et al. (2009). The paper gives details of the magnetism and the thermal history of this very important sample.

[6] Tikoo et al. (2012). This abstract presents the evidence for the decline of dynamo principally from work on sample 12022.

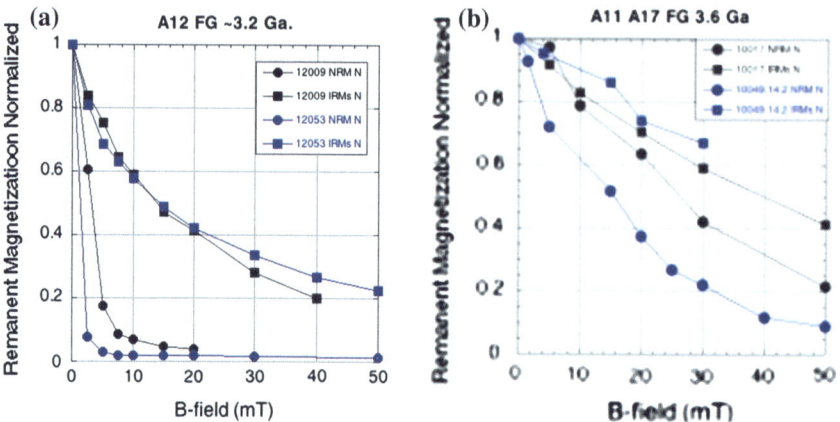

Fig. 9.6 AF demagnetization characteristics of normalized NRM and IRMs of **a** fine grained samples Apollo 12 samples and **b** fine grained Apollo 11 samples

normalized NRM and IRMs for petrologically fine grained Apollo 12 basalts for which data are available.

Note that the NRM is far softer than IRMs. It is also much softer than the NRM of two older Apollo 11 samples that appear to carry reliable primary NRM (Fig. 9.6b). Of course many more samples need to be analyzed before we can be confident of this interpretation and there are Apollo 11 samples that are very soft. Yet the distinct differences in the AF demagnetization fingerprints between the fine grained Apollo 11 and 12 AF demagnetization plots suggests that the former may indeed be carrying primary NRM recording an early lunar field, whereas the latter do not appear to be.

The group from Aix-en-Provence, consisting of Cécile Cournède, Jérôme Gattacceca and Pierre Rochette, has also provided evidence of stable NRM in additional lunar samples and has been able to get more paleointensity data. They presented the normalized AF stability curves and directional data to distinguish samples as we have done above. Out of 18 samples they selected 7 according to the criteria that (1) their NRM converged to the origin showing that their high microcoercivity fraction was in a single direction and probably primary and (2) their AF demagnetization behavior was similar to ARM the proxy for TRM. The results are shown in Fig. 9.7 for the 70017, 71505 and 71597 samples for which the comparison with ARM is most convincing. Note that these are normalized at 10, 20 and 5 mT, so as to minimize the effect of the very soft magnetization (Cournède et al. 2012).

70017 was also studied in the early Apollo days and had like 10020 given promising results. It has an age of 3.7 Ga and the new paleointensity result by the author's improved IRMs normalization gave ∼40 µT, which is in good agreement

9.3 Intensity of Recorded Fields from NRM 95

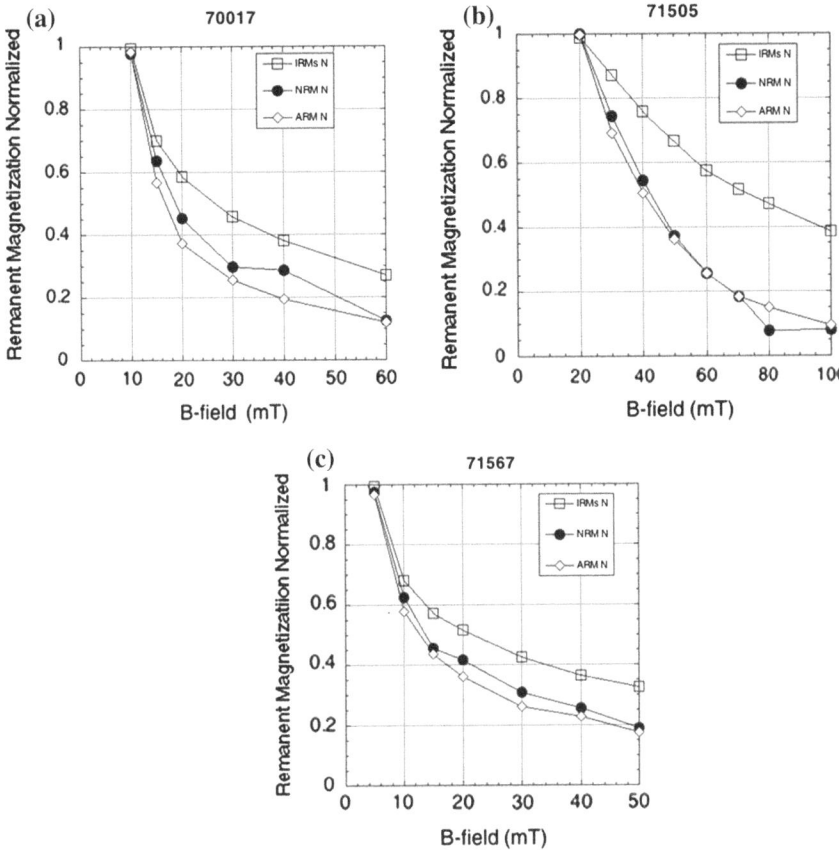

Fig. 9.7 Comparison of normalized AF demagnetization of IRMs, NRM, and ARM the TRM proxy

with the earlier work. No earlier work had been reported for 71505, which is a walnut sized rake sample. Neither is radiogenic age data available for this sample, although as a member of the high titanium low potassium group it may well have an age of 3.7–3.9 Ga. The high coercivity fraction trends to the origin and be primary and yields 90 µT. 71567 has a stable moment which survives to 48 mT AF demagnetization and again is origin trending and appears primary. The improved IRMs normalization method gives a value of 110 µT. Unfortunately, the age is unknown although there are indications that it may be one of the oldest mare basalts. The results for 70017 and 71055 have been added to the other results using the typical age for high Ti low potassium mare basalts (Fig. 9.7).

Figure 9.8 gives a summary of the available reliable results. The older samples all show similar results, but there is a major fall off by 3.2 Ga. For comparison, I have added an approximate value for the surface field of earth.

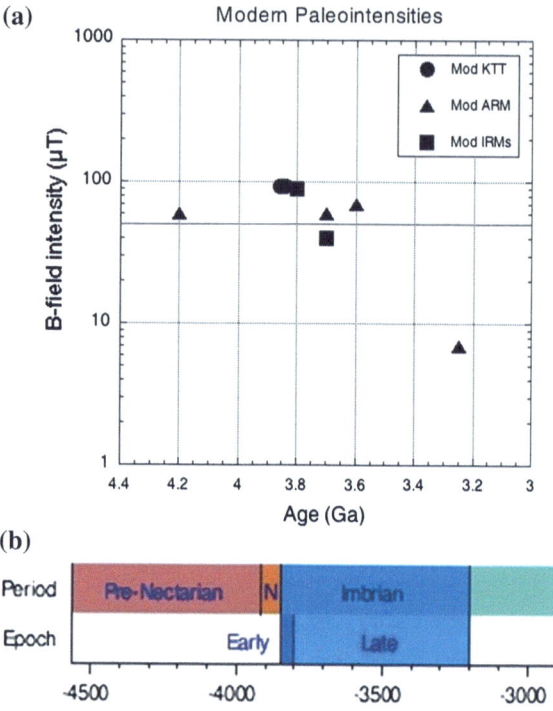

Fig. 9.8 a Recent lunar field paleointensity determinations by Koenigsberger-Thellier-Thellier, ARM and IRMs methods and **b** Oldest lunar stratigraphy

9.4 Paleomagnetism of Melt Breccias

The classical paleointensity results from the melt breccias play such an important role in these studies by establishing the maximum field around 3.9 Ga that we need to discuss their NRM in more detail. Important clues to the origin of the NRM of these rocks have come from the fieldwork carried out by the astronauts with the boulders at the Apollo 17 site. Figure 9.8 shows the Apollo 17 Boulder 1 at Station 2. This site provided one of the few occasions where we had multiple samples from a single melt breccia boulder. The astronauts with the fieldwork teams did a great job making a collection from what appeared to be different units within the boulder. The boulder had apparently rolled down from the nearby South Massif, which is visibly stratified. Some had even left tracks, which could be traced back to the site from which they had come. It appeared that we had a small sample of highland stratigraphy, probably within an ejecta blanket from the Serenitatis basin event (Fig. 9.9).

Subir Banerjee, an old friend from our Cambridge days, was working on the paleomagnetism of the samples from this boulder in conjunction with a consortium of geologists and geochemists. He and his associates were able to take advantage of the fieldwork on this boulder and the contributions of the consortium to give a detailed analysis of the NRM of the mutually oriented samples from the boulder

9.4 Paleomagnetism of Melt Breccias

Fig. 9.9 Boulder 1 at Apollo 17 station 2

that had not been possible elsewhere. An important clue to what is going on comes from the fact that boulder is layered and the magnetic behavior varies between layers. In your author's opinion the layering is correctly indicated towards the bottom left of the figure, but is incorrect at the top right. Given this, the layering is likely reflecting layering in an ejecta blanket (Fig. 9.9).

Banerjee and his colleagues studied numerous matrix samples from 72215, 72255 and 72275. Given the uncertainties, the NRM directions of the matrix subsamples are nearly indistinguishable, The similar directions of NRM contrasts markedly with the scattered directions of the higher coercivity magnetization in the matrix samples of 72255 and 72275 revealed by AF demagnetization. Hence, some event partially demagnetized the matrix layers and some clasts. However, this seemed to have taken place without fully overprinting a higher coercivity and higher blocking temperature fraction, which was likely a pre-assembly component. In contrast, the magnetization direction of the 72215 matrix was stable in AF demagnetization to 20 mT. 72215 was another sample from which the Scripps group got excellent paleointensity data (Fig. 9.7). Indeed the comparison between the 62235 and 72215 is a testament to their remarkable laboratory technique. When you realize that the two breccias are of identical age, the simplest conclusion is that they both cooled in ejecta blankets in very similar lunar fields of about 100 µT at a very similar time.

An obvious problem with the simplest conclusion is that we do not know how this NRM was acquired. We know from the petrology of these ejecta samples that the temperatures involved in their initial production are 1,200–1,500 °C. These are far higher than the Curie point of iron, so why don't the samples carry full TRMs that do not demagnetize until the Curie point, i.e. we need to account for the

Fig. 9.10 The boulder cluster at station 6 of Apollo 17

roughly 500 °C blocking temperature of most of the NRM. There is some magnetization above this temperature range, but it is very weak implying a very weak field. Clearly there are still problems here.

A similar study was carried out with the Boulder cluster at station 6 of Apollo 17. Figure 9.10.[7] Again excellent field data and petrological studies helped to show that this boulder cluster was likely a section through a single impact melt sheet. It formed at some distance from the impact point because there are clasts, which are only very weakly shocked (<10 GPa). Thus these clasts did not see shock induced temperature increases of as much as ∼200 °C and must have been 1,200 or 1,500 °C cooler than the typical impact melt in which they were engulfed.

The cluster we see was eventually formed by the breakup of a boulder that rolled down from a horizontal region of boulders, which may well have come from a single unit in the massif above. The interpretation of the boulder cluster history led Gose and colleagues to suggest that the NRM was composed of two components: a relic direction from clasts and a thermal overprint from final cooling in an ejecta blanket Gose et al. (1978). This interpretation is similar to that discussed above for the Apollo 17 Station 2 Boulder 1 study and may eventually lead to a satisfying explanation of the paleomagnetic record of melt breccias and melt rocks from ejecta blankets. An additional suggestion is the role of the size and number of clasts in the boulder in determining the local maximum temperatures reached. The detailed models of this need to be clarified quantitatively and it does appear that at least locally temperatures in the ejecta blanket must have been far above the Curie point of Fe, but the thermal imprint is an important suggestion relevant for many breccias. Above all we need thermal demagnetization of the NRM of samples from the Apollo 17 Station 6 Boulder cluster to test these ideas.

There is one terrestrial analogue that may be useful here in interpreting the NRM of melt rocks and melt breccias, even though it is obviously far from perfect. It is the hot cloud from volcanoes that can contain a mix varying in grain size from

---

[7] Courtesy NASA.

9.4 Paleomagnetism of Melt Breccias

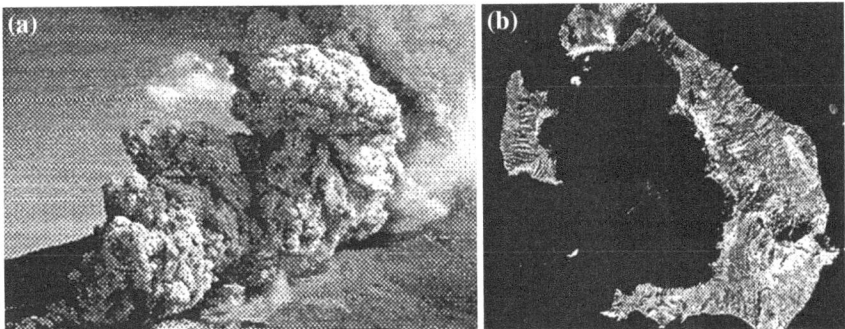

Fig. 9.11 a Mount St. Helen's eruption August 7, 1980 and **b** the crater left from the Santorini eruption

boulders to microscopic ash particles (Fig. 9.11a). These result in what geologists call pyroclastic deposits. The events at Mt. St Helens were well documented. The major event at Santorini was Late Bronze age at ~ 1636 BC and according to Plato wiped out a thriving civilization there (Fig. 9.11b). The effect on the home of the Minoan civilization is still debated, but it must have been a bad day on nearby Crete. Some of the pyroclastic deposits from the Santorini event have been studied by paleomagnetists, who used thermal demagnetization to find the temperature at which the major component of NRM demagnetizes. The temperature at which the NRM disappeared, i.e. was unblocked, was interpreted to be the temperature at which the deposit was laid down, so that below this temperature they acquired coherent NRM.

The method has had some successful general application to find temperatures of deposition of such deposits and the geomagnetic field at the time. Using the same approach with the Scripps data, it appears that the lunar ejecta blankets, in which 62235 and 72215 cooled, settled at about 500 °C. Following the interpretation of the results from the pyroclastic deposits, the low field recorded by the highest temperature NRM was probably either acquired in transit during minimal cooling, or is a ghost of the NRM, which survived from before excavation. In either case, it would presumably have been randomized by the event. Certainly, one would expect a very low field to be recorded. If this interpretation of these breccias holds up, other melt rock and breccias may also carry excellent records of the lunar field around 3.8–3.9 Ga and still others may record older fields at the time of major impact events. However, the low temperatures predicted by the pTRM in the ejecta blanket remain puzzling. Sometimes in science a very nice idea seems to fly in the face of some experimental data and yet ultimately turns out to be correct and the experimental data can be interpretable consistently. This may ultimately be shown to be the case here, so for the moment let us accept and follow the paleomagnetic interpretation of the terrestrial analogy for the lunar melt breccias.

9.5 Paleointensity Summary

Returning now to Fig. 9.8, we find a field of around 100 µT at ∼3.9 Ga from the new data, the IRMs normalization results (Fig. 9.4), the intensities from standard methods from the Apollo days (Fig. 9.5), and the recent results from 62235 and 72215. Therefore one can have some confidence in this result. Prior to ∼3.9 Ga, there is only one modern determination and that indicates a similar but smaller field at 4.2 Ga. The older Apollo era data suggest a decrease in field intensity, but there are few reliable results. After ∼3.9 Ga there are modern results from 10020, 10017 and 10049 that suggest a possible minor decrease in intensity by 3.7 Ga, with a major decrease shown by the analysis of 12022 at 3.2 Ga. The simplest interpretation is that a lunar dynamo turned on within a few hundred million years of the formation of the moon, but shut off between about 3.2 and 3.5 Ga. However, with so little reliable data, it is not only important to get more results, but to test this model independently.

9.6 Crustal Magnetism: Magnetic Anomalies

The crustal magnetic anomalies afford, in principle, a data set against which the dynamo model interpreted from paleomagnetism of the samples may be tested. By way of introduction it should be said that magnetic anomalies are the departure from the regional field caused by local magnetic sources. They have been a major part of geophysical exploration of the earth's crust both for magnetic materials, such as iron ore, and as a means of establishing structure and depth of basins in oil exploration. They also played a key role in establishing the veracity of sea floor spreading (Chap. 6). These applications have given rise to sophisticated analyses of anomalies to interpret the magnetism of the source rocks.

Remembering that the magnetic maps of many terrestrial craters show a strong central anomaly within a region of subdued magnetically subdued region, a possible test for an ancient lunar field would be to look for similar anomalies magnetic in the center of large craters on the moon. Halekas and colleagues[8] did precisely this and found that the multi-ring basins in general exhibited a magnetic low, which in some cases included a central anomaly. They found that central anomalies were strongest in early Nectarian basins and decreased in Imbrium age basins, and were absent in the oldest Pre-Nectarian basins. These results confirm

---

[8] Halekas et al. (2003). This paper gives a list of more than 30 lunar impact basins and the nature of their magnetic anomalies, in terms of the presence or absence of crater lows and central magnetic anomalies.

9.6 Crustal Magnetism: Magnetic Anomalies

the results of an ancient lunar magnetic field, but the timing is not consistent in detail with the results from the samples. Analysis by Hood of surveys of 4 Nectarian age multi-ring basins again demonstrated associated central anomalies.[9]

South Pole Aitken (SPA) is the oldest and largest of the lunar basins. With the new data from Lunar Prospector yielding detailed surveys of South Pole-Aitken feature, efforts to interpret its magnetic anomalies led to two important recent papers. The first by Wieczorek, Weiss and Stewart[10] argued that the anomalies were so strong that they could not have originated from endogenous lunar material. Rather they advocated that highly magnetic material had come from the impacter. The second paper by Purucker, Head and Wilson[11] focused upon the lava pools within SPA and the feeder dykes to them. They had noticed that the anomalies over much of the area had linear, or gently arcuate trends and that some of the stronger anomalies were associated with some of the larger lava pools. They then proposed a model of dykes, which had fed the lava pools and had the tabular form that could give rise to the linear anomalies seen. The dykes that caused the linear anomalies would be like the dyke swarms on earth with individual dykes being blade like, with lengths of tens to hundreds of km and widths of tens to hundreds of meters and occurring in groups 10's of km wide. These two papers make different predictions for the age of the source rock for the anomalies. If the source is the ejecta, then the age of the magnetization will be similar to the age of SPA. However, if the source is the dyke system feeding the lava pools, this is likely to be significantly younger. Thus if the ejecta model is correct then we have additional evidence for a dynamo field at ~ 4.3 Ga, but if the dyke model is correct the required magnetic field could be much later.

How well does the early dynamo hold up in the tests from the crustal anomalies? The magnetic anomalies require strong magnetic source rocks. They therefore confirm the primary conclusion of the need for a dynamo field in a lunar core. They do provide confirmation for a lunar core dynamo functioning around $\sim 3.7-3.9$ Ga. Recent work on the magnetic anomalies suggest that the strongest anomalies are predominantly due to ejecta of extralunar impact material. Moreover, analysis of the anomalies indicates that they record dipole fields of both polarities indicating a field reversal.[12]

---

[9] Hood (2011). In this paper Hood provides exquisitely detailed magnetic anomaly maps for four major Nectarian aged basins.

[10] Wieczorek et al. (2012). This paper develops a model for the magnetic anomalies of South Pole Aitken depending upon material from the impacter to account for their strength.

[11] Purucker et al. (2012). This paper follows earlier work linking dykes Swarms to the lava poles in South Pole-Aitken and interpreting the magnetic anomalies by massive feeder dykes.

[12] Private communication Mark Wieczoreck. Many thanks.

9.7 Summary

We conclude that there was an early lunar dynamo, but exactly when it turned on and off is less certain. An obvious difficulty with assessing the history of the lunar dynamo arises from the very small number of successful paleointensity estimates, but we are in the middle of an effort to get more. Now let's see in our final chapter how an ancient lunar dynamo field might fit into the grand scheme of lunar history.

References

Cournède C, Gattacceca J, Rochete P (2012) Magnetic study of large Apollo samples: possible evidence for an ancient centered dipolar field on the moon. EPSL 331–332, 31–42
Collinson DW, Stephenson A, Runcorn SK (1973) Magnetic properties of Apollo 15 and 16 rocks. Lunar Planetary Science Conference 4, vol. 3, pp 2963–2976
Garrick-Bethell I, Weiss BP, Shuster DL, Buz J (2009) Early lunar magnetism. Science 323:356–359
Gose WA, Strangway DW, Pearce GW (1978) Origin of magnetization in lunar breccias—an example of thermal overprinting. Earth Planet Sci Lett 38:373
Halekas JS, Lin RP, Mitchell DL (2003) Magnetic fields of multi-ring impact basins. Meteorit Planet Sci 38(4):565–578
Hood LL (2011) Central magnetic anomalies of Nectarian-aged lunar impact basins: probable evidence for an early core dynamo. Icarus 211:1109–1128
Lawrence K, Johnson C, Tauxe L, Gee J (2008) Lunar paleointensity measurements: implications for lunar magnetic evolution. Phys Earth Planetary Interiors 168:71–87
Purucker ME, Head III JW, Wilson L (2012) Magnetic signature of the South Pole-Aitken basin: character, origin, and age. J Geophys Res 117:EO5001. doi:1029/2011JE003922
Shea EK, Weiss BP, Cassata WR, Shuster DL, Tikoo SM (2012) A long-lived lunar core dynamo. Science 335(6067):453–456
Suavet C, Weiss BP, Fuller M, Gattacceca J, Grove TL, Shuster DL (2012) Persistence of the lunar dynamo until 3.6 billion years ago, LPSC 43, 1925
Sugiura N, Strangway DW (1983) Magnetic Paleointensity determination on lunar sample 62235, Lunar Planetary Science Conference 13, A684-A690
Tikoo SM, Weiss BP, Grove TL, Fuller MD (2012) Decline of the ancient lunar core dynamo, 43rd LPSC., ABS, 2691
Wieczorek MA, Weiss BP, Stewart ST (2012) An impactor origin for lunar magnetic anomalies. Science 235:1212–1215

Chapter 10
Lunar Magnetism in the Grand Scheme of Lunar History

A key result from the Apollo program was an acceptable model for the origin of the moon, in the form of a collision with Earth of a Mars size impacter. The moon then forms from the material remaining in orbit around thy earth. This model cast some doubt about the presence in the moon of enough iron for the core needed for a lunar dynamo, but it does permit a small core. Recently, a core of about 300 km has been "seen" by two groups of seismologists. Moreover, there is a solid inner core surrounded by a fluid outer core similar to Earth, but on a much smaller scale. Questions remain concerning the energy source to maintain a dynamo for nearly a billion years in so small a core.

10.1 The Kona Conference and a Model of the Moon's Origin

A decade after the completion of the manned missions to the moon a conference was planned by the Lunar and Planetary Institute to review new ideas about the moon from Apollo.[1] It was held in Kailua-Kona, on the Big Island, as Hawaii is known locally. The meeting is best remembered for the general recognition and acceptance of the idea of the giant impact origin of the moon. As is so often the case, there had been precursors. Papers advocating the idea had appeared from Hartmann and Davis and from Cameron and Ward in 1975. As we learnt in Chap. 2 , there were earlier antecedents in the C19th and therewas also the paper by Reginald Daly in the 1940s. The notion was that earth suffered an impact of a Mars size object and that subsequently the moon had formed from the ejecta that had remained in earth orbit. Given an early solar system in which the formation of the terrestrial planets was held to be by successive collisions of planetesimals, the late impact proposed seemed to fit naturally into generally accepted ideas. The Mars sized object was appropriately named Theia, who had given birth to Selene, the

---

[1] Hartmann et al. (1986). This book provides the proceedings including key papers on the origin of the Moon.

admirer of Endymion. Not everyone was convinced, but the main focus was now on how this might have taken place and how the material left in orbit about the earth gave rise to the moon.

You may remember from the introduction how difficult models of the origin of the moon had proved back in the 1960s. For each idea, there seemed insurmountable problems. The Apollo samples made at least two of them even less tenable. The venerable fission theory held that the moon had spun out of the earth during an early period of high rotation. However, the Apollo samples showed that the rocks were sufficiently different from earth's rocks to invalidate this idea. A second idea was the simultaneous origin of the earth and moon, as a twin planet system originating by condensation from the same solar system cloud. It too faced problems in the compositional differences between the earth and moon. The third possibility was an origin of the moon elsewhere in the solar system and subsequent capture by earth, but the odds against capture are unacceptable. Over any model of the origin of the moon hung the problem that the earth moon system's angular momentum was so large.

The giant impact theory provided just what was needed. Figure 10.1 shows the first stage of an oblique impact of the Mars size impacter. After contact a jet of hot vaporized rock squirts from the impact point on earth. This does not include high density core materials of the two bodies. It eventually cools and provides the material to form the moon, immediately accounting for its low density. However, the ejected material included enough of the earth that the bulk composition of the earth and moon would be similar, as is now evident. The principal difference would be the lack of volatiles, which would have condensed in space and been lost, so that the model predicts that the moon would be very dry, as indeed it is. Later model simulations also showed that the impact could leave the earth moon system with the appropriate angular momentum.

Let us assume that the basic idea of the giant impact origin of the moon is correct, and see how lunar magnetism might fit into such a scenario. Immediately there is a problem. The initial model predicted that there should be no metallic

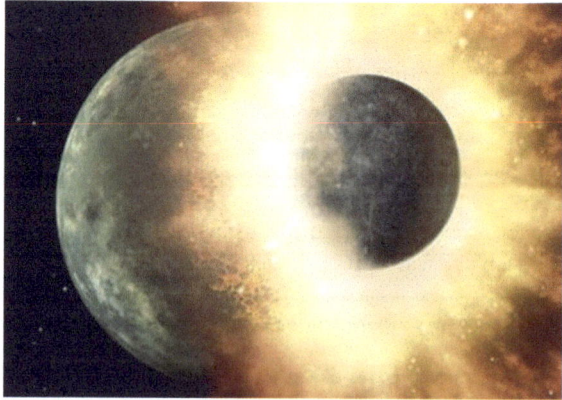

Fig. 10.1 Giant impact theory of origin of the Moon (Courtesy NASA/JPL Caltech)

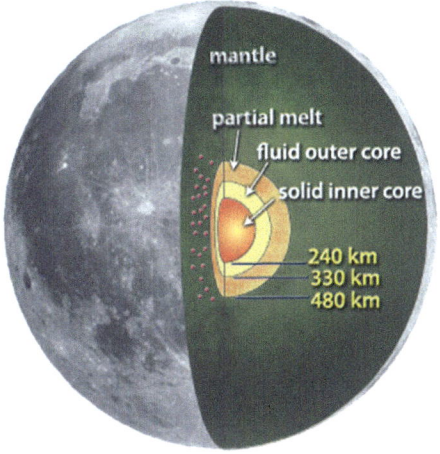

Fig. 10.2 Lunar interior showing molten outer core [Courtesy NASA and special thanks to Renee Weber, see also Weber et al. (2011), Garcia et al. (2010)]

core in the moon, or at most a very small one. That is obviously not good for any model of lunar magnetism, which requires a molten core in which to generate any lunar dynamo field. Recent seismic studies have however saved the day for us by detecting a lunar core.[2] The paper by Garcia and colleagues is entitled "Very Preliminary reference Moon model". You may again remember that the lunar seismic signals (Fig. 5.3) did not look very promising for the determination of the lunar interior by the techniques used on Earth. This title reflects the enormity of these difficulties even with advances in 40 years since Apollo. However, remarkable new analyses by the two groups have given consistent evidence of a core. Garcia and his colleagues give a core radius of 380 ± 40 km and Weber and her colleagues give between ∼250 and 430 km. Both groups also determine a molten outer core at present and give evidence for a solid inner core. The Weber paper suggests that the small temperature difference between the Inner core boundary and the core mantle boundary would give a stably stratified fluid core rather than an actively convecting core. This permits the same mechanisms for dynamo action as in the Earth at an earlier time and explains why there is no field at present even though there is a partly molten core (Fig. 10.2).

To set the stage for the early solar system in which the moon originated, we turn to the isotope geochemists with their ability to date events with exquisite accuracy. Key roles were played by short lived isotopes, with half lives of a few million years or less, e.g., ^{26}Al and ^{182}Hf. The former provided the heat source that drove the evolution of bodies in the early solar system and the short half lives of both made possible age determination with high resolution. The questions being settled are the sequence and ages of the earliest events in the solar system and the

---

[2] Courtesy NASA and special thanks to Renee Weber, see also Weber et al. (2011), Garcia et al. (2010).

precision of their timing seems to me a truly remarkable achievement, awesome as our younger readers might say!

Our clues to this history come from the study of meteorites.[3] They can be conveniently divided into irons that mostly come from the cores of differentiated parent bodies, stones that come from the crust/mantle of such bodies and from bodies that never melted, and stony irons that appear to sample both core and crust/mantle material. The first objects that condensed in the solar system were the Calcium-Aluminium-Inclusions (CAI's). They are found in primitive stones and have been dated at 4.567 Ga. They reveal that the nebular gas and dust was then at a temperature around 1,200–1,400 °C. Another feature of some primitive stones is the presence of chondrules—small silicate spherical droplets that give the name chondrites to the meteorites containing them. Initially they too were thought to have come from the nebular gas like the CAI's.

If we take the age of the CAI's as zero, then by the age of 2–5 Ma, differentiated asteroids with molten cores had formed and their cores were sampled by the oldest iron meteorites. Within this first one or two million years, planetesimals had formed by accretion and melted to give cores. Collisions between differentiated plantesimals, or asteroids, produced samples of cores and mantles or crusts, in the form of iron, stony-iron, and non-primitive stone meteorites. It is not clear in this new emerging paradigm whether chondrules were initially formed in the solar system nebula from aggregates of solid particles that were then melted, as was previously thought, or whether they were products of collisions between planetesimals (Scott 2007). The collision that formed the moon took place some tens of millions of years after the zero from the CAI's, say at about 4.53 Ga.

There is one final aspect of lunar history that may play a key role in lunar magnetism. This is known as the Late Heavy Bombardment. The suggestion is that at about 3.9 Ga the moon suffered a peak in its bombardment history reflected in the age of huge impact basins, of many of the Apollo and Luna returned melt rocks and melt breccias and of many of the lunar meteorites found in Antarctica. Recently confirmation of the idea of a Late Heavy Bombardment has come from a solution to a very different problem. Dynamic models of the early solar system were being generated by a group of theorists that included Tsiganis, Morbidelli, Gomez and Levison and were presented in a paper at the Lunar Science Institute in 2005 by Harold Levison and Bill Bottke (6). The essence of the idea is that the present configuration of the outer planets is not the original, but that rather they were initially in a smaller region, in which their gravitational interactions

---

[3] There are several "popular" books on Meteorites such as *Falling stars: a guide to Meteors and Metorites,* 2nd Edition. Mike D. Reynolds, 2010, Stackpole books, Meteorites by Smith, C., Russell, S., and Benedix, G., 2010, Firefly Books Ltd., *Meteorites and their Parent Bodies,* McSween, 1999, Cambridge University Press is a little more technical, but very entertaining. A technical reference for early solar system matters is *The Treatise on Geochemistry Volume 1, Meteorites, Comets and Planets,* 2004, Ed. Davis A.M. It contains review papers on the Classification of Meteorites, Chondrules, the Origin and Earliest history of the Earth, Early Solar system chronology, and the Moon.

destabilized their orbits. In the ensuing spreading out of the orbits of the planets, the icy remnants of the early solar system were disturbed, and the inner solar system suffered intense bombardment with this debris around 3.9 Ga. Of course this is precisely when the strongest magnetic fields are suggested by the lunar paleointensity results.

10.2 Models of the Lunar Magnetism

Given a model of the origin of the moon by a giant collision, the moon must have formed with an orbit that was much smaller than at present. There are suggestions that the orbit may also have been far more elliptical than it is now and that during this time the near side far side asymmetry of the moon was initiated. Like the earth, it must have experienced a period of cooling and solidification from a molten mass. During this time iron could have migrated to form a core. If the paleomagnetic results described in the last chapter are correct, we need a lunar core that was partly molten at 4.2 Ga, a few hundred million years after the formation of the moon. In addition, we have evidence that such fields persisted on the moon for several hundreds of millions of years (Fig. 9.7). Next, we ask how it might have operated.

There are two basic ways that magnetic fields can be generated in molten metallic cores of planets, moons and possibly asteroids. In our discussion of the origin of the magnetic field of the earth, we assumed that simple thermal convection or the ascent of low density fluid arising from the formation of the inner core would generate the motion required to maintain a dynamo. Since the moon has a smaller scale version of the same core, the energy sources that give the geodynamo might have operated in the moon at some point (Konrad and Spohn 1997). The puzzle is how long could so small a core give rise to a dynamo like the earth's by active convection, before settling into the state we see the moon in at present with a partly molten core with no magnetic field.

The second way for a dynamo to operate relies on a mechanism, such as precession or tidal deformation, to give relative motion between the molten core and the solid mantle, which then generates flow in the outer core that can drive dynamo action. At various times, it has been suggested that these mechanisms could drive the terrestrial dynamo. However, the conventional wisdom has always favored convective models, or buoyancy in fluid remaining from inner core formation to drive the dynamo. Early objections to precession as a source for the geodynamo were that there was insufficient energy from the precession to drive a dynamo and that the flows generated by this mechanism were not suitable for dynamo action. However, Jim Vanyo at UCSB and others had long argued that precession could provide the necessary energy source and suitable flows for dynamo action. Working with Bob Dunn, Jim had carried out experiments, which demonstrated his ideas. The experiment utilized a spheroid filled with water to represent the core. This was placed on the spin table, which was rotated about an

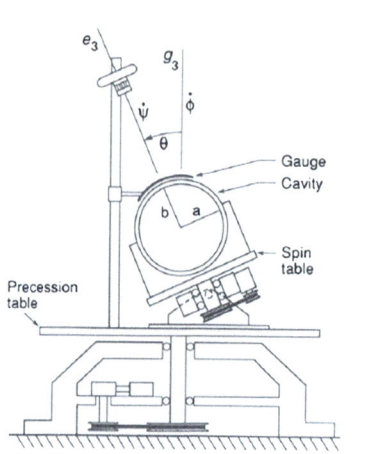 

Fig. 10.3 a (*Left*) spinning inclined table placed on rotating table to simulate precession of rotating earth, and **b** (*right*) pattern of flows generated. *White spots* with *blue haloes* are artifacts from illumination lights

inclined axis. The spin table was itself rotated by the precession table, with the ratio of the rotations corresponding to the ratio of the earth's rotation to precession (Fig. 10.3a). Admittedly the fluid used was water, of which the core is certainly not made. Nevertheless, as Fig. 10.3b[4] shows there were flows, which could well generate shears like those required for dynamo models (Appendix 5).

Other work had also demonstrated that plenty of energy was available from the process and the debate would turn on whether nature had utilized this energy to power the dynamo. Some paleomagnetic evidence for a role of both precession and tidal forces on earth has appeared, but is not generally accepted. The need for such energy sources was not evident for the geomagnetic dynamo because between the convective cooling of the molten core, the latent heat from the formation of the inner core and the buoyant low density material liberated from the formation of the inner core, adequate energy sources were thought to be available. In the lunar case, something other than convection was much more attractive because the dynamo appears to have functioned in a very small core for nearly a billion years (Fig. 9.7). Yet, the heat flow from the core decreases sufficiently during this time to preclude further cooling by convection. A dynamo might have operated in the lunar core for some of the time in the first billion years. Nevertheless, the problem of the longevity of a convection driven lunar dynamo in so small a core is

---

[4] Courtesy Jim Vanyo and Bob Dunn. See also Vanyo (1991), Vanyo and Dunn (2000). These papers show that flows with sufficient complexity can be generated by precession to make dynamo action feasible. There is a far larger literature investigating precession as a possible energy source for dynamo action.

10.2 Models of the Lunar Magnetism

generally recognized as serious and suggestions for other power sources for the early lunar dynamo have been made (Dwyer et al. 2011; Le Bars et al. 2011).

One, by Dwyer, Stevenson and Nimmo follows the ideas on precession as a driving force, but adds an intriguing wrinkle based on the orbital history of the moon. They note that early in lunar history the moon was much closer to the earth than it is now and the energy available for a precession driven dynamo would be correspondingly larger. As the moon receded from earth, the precession power source would decrease and eventually the dynamo would fail. Such a beautifully simple idea, which explains the observations so well, makes a powerful case. Presumably a similar argument might be made for the eventual failure of a tidal deformation mechanism.

The second paper, by Le Bars and colleagues, faces the same difficulty of an alternate power source to convection. They argue that the giant impacts were sufficient to change the rotation state of the moon and that because the changes in the mantle and the mantle and core would differ, relative motion would be set up between the core and mantle. Given this, they argue that flows suitable for dynamo action would follow. They also propose that the impacts would take the moon out of lock with the earth and that tidal dissipation would add further power.

10.3 Summary

So how do lunar paleomagnetism and lunar crustal magnetism fit into the grand scheme of lunar history? The melt rocks and melt breccias as they cooled in ejecta blankets appear to record the strongest magnetic fields at an age of ~ 3.9 Ga. These melt rocks are clearly stronger sources of lunar magnetism than the mare basalts. However, the mare basalts suggest that the dynamo lasted beyond the age of about 3.9 Ga, perhaps to near Apollo 12 basalt times ~ 3.2 Ga. Moreover, the troctolite appears to carry a primary NRM recording a field from 4.2 Ga. Although some argue that a convection driven dynamo could run in the proposed lunar core for several hundred billion years, another energy source may be needed. This source is going to come one way or another from some form of mechanical stirring giving rise to relative motion on the core mantle boundary. Whether this will be directly impact driven, or by a precession, or tidal model is not yet clear, although the simplicity of the precession model is very appealing. This is not a time to be too dogmatic about dynamo mechanisms in the moon. It is an area of active research with new analyses and ideas emerging. What is clear from lunar paleomagnetism and lunar crustal magnetism is that the moon had a molten core, in which a dynamo gave lunar fields of tens of μT, similar to the magnitude of the present surface field of the earth, and that they may have lasted for near to a billion years.

References

Dwyer CA, Stevenson DJ, Nimmo F (2011) A long-lived lunar dynamo driven by continuous mechanical stirring. Nature 479:212–214

Garcia RF, Gagnepain-Beyneix J, Chevrot S, Logonn P (2010) Very preliminary reference moon model. Phys Earth Planet Inter 188:96–113

Hartmann WK, Phillips RJ, Taylor GJ (eds) 1986 Origin of the moon

Konrad W, Spohn T (1997) Thermal history of the Moon: implications for an early core dynamo and post-accretional magmatism. Adv Space Res 19(10):1511–1521

Le Bars M, Wieczorek MA, Kartekin O, Cebron D, Laneuville M (2011) An impact-driven dynamo for the early Moon. Nature 479:215–217

Scott ERD (2007) Chondrites and the Protoplanetary Disk. Ann Rev Earth Planet Sci 35:577–620 (This provides an exhaustive discussion of Chondrules, Chondrites and the early solar system)

Vanyo JP (1991) A geodynamo powered by luni-solar precession. Geophys Astrophys Fluid Dyn 59:209–234

Vanyo JP, Dunn JR (2000) Core precession: flow structures and energy. Geophys J Int 142:409–425

Weber RC, Lin P-Y, Garenro EJ, Williams Q, Logonné P (2011) Seismic detection of the lunar core. Science 331:309–312

Appendix A
Rotational Mechanics Background

There are two great conservation principles that are fundamental to an understanding of much of physics and we will use them several times in this book.

Energy and its conservation

The conservation of energy tells us that energy may change form, but it is not destroyed, or created. Technically energy is the capability of a system to do work. Its conservation can be demonstrated by the example of a swinging pendulum. The total mechanical energy of a pendulum is the sum of its potential energy (PE), the energy due to its position and its kinetic energy (KE), due to its motion (Fig. A1.1).

When the pendulum mass is at its maximum departure from rest it has its maximum potential energy (PE = mgh), where g is the gravitational constant. It then experiences a force (F = gh). As it moves towards the low point, PE is decreasing because h is decreasing. Meanwhile, the kinetic energy (KE = $^1/_2$ mv^2), is increasing because v is increasing. The KE reaches a maximum at the low point when the PE is at a minimum. As the pendulum continues to swing the exchange of energy continues until frictional losses eventually stop it.

Another critical aspect of energy, which we will encounter in magnetism, will be the idea of energy minimization of a system. This will take many different forms, but the overall notion is that a particle, or a system of particles, will tend towards a minimum energy state. For example, a compass needle is in a lower energy state when it is parallel to the field and hence it rotates towards that direction.

Still another aspect, which we will encounter, is the role of thermal energy with its generation of disorder. For example, magnetism is due to exchange energy interactions between atoms, which align their individual moments. This alignment dominates at room temperature in the materials we call magnets. However, if the material is heated, it will reach a temperature at which the disordering effect of thermal energy will overcome the ordering due to the exchange energy and the material will cease to be a magnet. Perhaps you saw this demonstrated at high school by heating a screwdriver with nails attached. As this was heated, eventually the nails fell off because they had ceased to be magnetic. Thermal energy had won again.

M. Fuller, *Our Beautiful Moon and its Mysterious Magnetism*,
SpringerBriefs in Earth Sciences, DOI: 10.1007/978-3-319-00278-1,
© The Author(s) 2014

Fig. A1.1 Conservation of energy in a pendulum

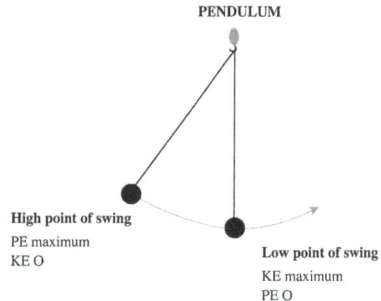

Competition between order and disorder is universal and not just in magnets. Moreover, the temperature at which disorder wins varies over a huge range. For example, superconductivity which we will encounter in the operation of superconducting magnetometers is governed by much weaker forces, which are overcome in the materials we use a few degrees above absolute zero.

Angular momentum and its conservation

To understand the conservation of angular momentum, remember the frequently cited example of the skater, who increases her rotation speed by pulling her arms in towards her body (Fig. A1.2).

Fig. A1.2 The conservation of angular momentum used by Yuko Kawaguti, when she pulls her arms close to her body from an extended position. By reducing her moment of inertia, she increases her rotation speed (Cup of Russia. Courtesy of creative commons)

Appendix A: Rotational Mechanics Background

To explain this, we need to clarify the fundamental concepts. The momentum of a body is the product of it mass and velocity.

$$\text{Momentum (p)} = mv$$

For a point mass rotating about an axis, its moment of inertia (I) is the product of its mass and the perpendicular distance from the rotation axis (r) squared, or

$$I = m\,r^2.$$

For a body made of many particles the moment of inertia is obtained by adding up all of the products of the particles that make up the body and their perpendicular distance from the rotation axis squared.

The angular momentum (L) of the body is the product of the moment of inertia and its speed of rotation (ω), or

$$\text{Angular momentum (L)} = I\omega.$$

Returning to our skater, she reduces her moment of inertia (I) by bringing her arms closer to her rotation axis. In doing so the conservation of angular momentum ensures that her speed of rotation (ω) must increase.

What can change the angular momentum (L)? Just as a force can change the momentum of a body, a torque (τ) can change the angular momentum of a body by changing ω. Clearly a torque must involve a force somehow. When you push open a door, you are applying a torque: you are applying a force (F) perpendicular to the door's surface at a distance from the hinges. The further from the hinges you apply the force the greater the effect, so the distance from the hinges must be involved. In fact, we can define a torque as the product of a force (F) and the perpendicular distance to the rotation axis (r),

$$\tau = Fr.$$

Now that we have introduced these two great conservation principles and some basic ideas of rotational mechanics, we are ready to return to our story.

Appendix B
Units

Metric prefixes

giga (G)	10^9	1 billion
mega (M)	10^6	1 million
kilo (k)	10^3	1 thousand
milli (m)	10^{-3}	1 thousandth
micro (μ)	10^{-6}	1 millionth
nano (n)	10^{-9}	1 billionth

Examples of magnetic units we will need

Much of the trouble with magnetic units is that the older centimeter gram second (cgs) system has only given way slowly to the SI Meter Kilogram Second (MKS) system, with the progressive demise of folks like myself brought up on Centimeter Grams Seconds (cgs). Here, we will give useful equivalents as a means of avoiding too much discussion of units.

Magnetic fields

B field	The unit is the Tesla named for Nikola Tesla the Serbian American physicist, among whose many contributions was the development of the alternating current electrical power supply system. The Tesla is defined in terms of the Lorentz force, which is experienced by a charged particle in motion in a B field. We met the Lorentz force in discussing the Van Allen radiation belts. The tesla is a rather large unit and we will most often encounter milliTeslas and microTeslas, or even nanoTeslas.
T	Some of the strongest magnetic fields from electromagnets, and those used medically in MRI are of the order of Teslas.
mT	milliTesla 100 s mT AF demagnetizing fields used in paleomagnetic analysis.
μT	microTesla $\sim$10 s μT geomagnetic surface field.
nT	nanoTesla which is equal to the old CGS unit the gamma—γ

Magnetic moments and magnetization

Am^2 Ampere meter^2 Magnetic moment—without going too far into electromagnetism, let us think about the moment of a magnet in terms of the torque a magnet experiences in a magnetic field. Thus, the torque that a compass needle experiences in the earth's field to turn it parallel to the field depends upon the magnetic moment of the needle and the strength of the field.

Am^{-1} Magnetic moment per unit volume. (Am^2/m^3 or Am^{-1}) magnetization or Magnetic moment per unit mass, which will be Am^2/kg.

Sometimes magnetization is reported as moment, but it is probably better to report as magnetic moment/unit volume, or/unit mass. Then one can compare the remanent magnetization of different samples more easily.

The unit of magnetic moment per unit mass has the advantage of being the same in both systems.

The most strongly magnetized lunar rocks tend to be regolith breccias, melt breccias and melt rocks, with intensities of 10^{-4} Am^2/kg, while mare basalts tend to be weaker with intensities of 10^{-6} Am^2/kg. There are however notable exceptions from these values.

Appendix C
The Alphabet Soup of Remanent Magnetisms

NRM	Natural Remanent Magnetism—carried by a rock in Nature.
IRM	Isothermal Remanent Magnetism—acquired on simple exposure to a magnetic field.
IRMs	Saturation Isothermal Remanent Magnetism—acquired after exposure to saturation magnetic field (Sometimes SIRM, sometimes Mrs, sometime Mr.).
TRM	Thermal Remanent Magnetism—acquired on cooling in a magnetic field from the Curie point to ambient temperature
pTRM	partial Thermal Remanent Magnetism acquired on cooling in a magnetic field in a temperature interval less than the temperature from the Curie point to ambient temperature.
PRM	Piezoremanent Magnetism—acquired on exposure to stress in a magnetic field.
SRM	Shock Remanent Magnetism—acquired on exposure to shock in a magnetic field.
CRM	Chemical Remanent Magnetism—acquired with formation of magnetic particles in a magnetic field.
VRM	Viscous Remanent Magnetism—acquired on exposure over extended time in a magnetic field.
ARM	Anhysteretic Remanent Magnetism—acquired when ramping down an Alternating Field, as in AF demagnetization, but in the presence of a magnetic field.
N.B.	The spelling remanent is not the usual spelling in the US, which is remnant, but this archaic or British form is retained in our discussions of magnetism. It will be interesting to see how long it lasts before giving way to the standard spelling.

Appendix D
Magnetometers for Sample Measurement

The standard instrument in Apollo time was the venerable spinner magnetometer, which appeared in various forms of which probably the most popular was the Schonstedt spinner shown in Fig. A4.1. The sample was mounted in the cube holder seen to the left in the figure. To measure, the assembly was moved along the track taking the sample into the magnetic shields, within which were pick up coils mounted perpendicular to the spin axis. As the sample spins, the magnetic field in the pick up coils changes and so in accordance again with Faraday's law a voltage is induced in them. The intensity of the voltage depends upon the intensity of the component of remanent magnetization in the plane of the coils and its phase relative to the axes of the system is dependent upon the direction of remanent magnetization. By analyzing this signal, the direction and intensity of the component of remanent magnetization in the plane of the coils can be recovered. By repeating the measurement along the other axes of the sample the total remanent magnetization can be obtained.

Just before the Apollo days, a new instrument had been introduced into the paleomagnetic community by Bill Goree and his colleague Bill Goodman, who

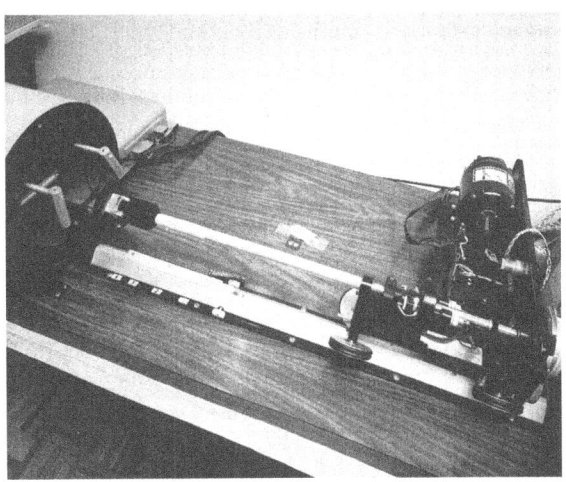

Fig. A4.1 Schonstedt spinner magnetometer of the Apollo era

M. Fuller, *Our Beautiful Moon and its Mysterious Magnetism*,
SpringerBriefs in Earth Sciences, DOI: 10.1007/978-3-319-00278-1,
© The Author(s) 2014

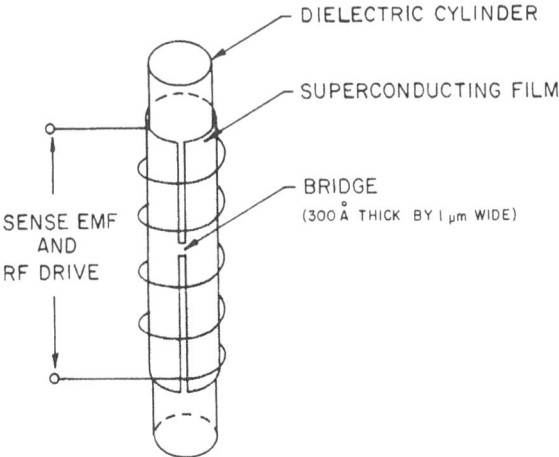

Fig. A4.2 The superconducting quantum interference device sensor (SQUID)

together in appropriate Silicon Valley style had started their 2G Company in Bill Goree's garage.

For standard measurements the sensor gave promise of rapid measurements at high sensitivity. However, because of the way the sensor worked, it became clear that there were many other possibilities, so let's look to see how the sensor works (Fig. A4.2). The sensor is fed by a superconducting circuit from the pick up coils into which the sample is introduced. Note that this current is flowing in a superconductor, so therefore it does not decay on the time scales, which concern us. At the sensor, the coil wound around the superconducting film generates a magnetic field. As prescribed by Faraday the change in magnetic field gives a current flowing to oppose this change. The design of the sensor with the small bridge carrying all of the current gives an effective amplification and allows one to take advantage of the fact that the type of superconductor used fails at a certain critical current and returns to a normal state. The design of the bridge ensures that the superconductor will reach its critical current at this point. When the critical current is reached the superconducting film fails, a flux quantum enters the cylinder and the sensor resets. One can count the flux jumps, which with the remaining current in the sensor gives a measure of the magnetization of the sample. As usual the electronics work a little differently, but the insertion of the sample does give a current in the film that is used as the measure of its remanent magnetization. Moreover, any change of the magnetization of the sample can be measured absolutely. Such a system is illustrated in Fig. A4.3.

This operation of the magnetometer immediately raised all sorts of possibilities. If one wanted to do the thermal demagnetization (discussed in Chap. 7), one could measure the changing magnetization as the sample heated up. Obviously other methods would be possible such as seeing directly the effect of stress or shock. Clearly not only the sensitivity, but the versatility of the instrument would make it an invaluable asset for paleomagnetists.

Appendix D: Magnetometers for Sample Measurement

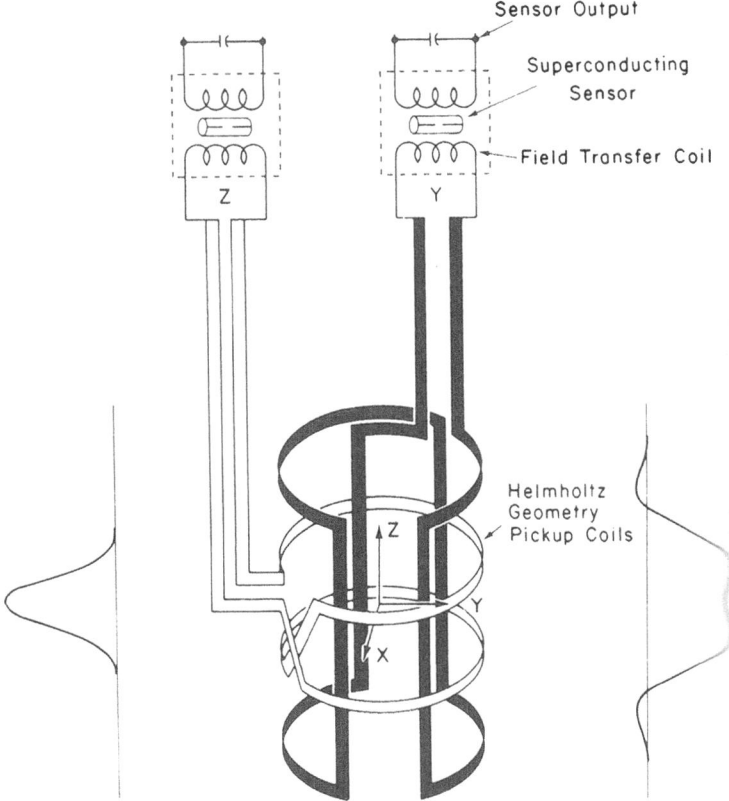

Fig. A4.3 A SQUID magnetometer with the output of the coils shown as a sample is passed through the pick up coils

Note

1. Goree, W.S., 2007, *Rock Magnetometer, Superconducting*, In Encyclopedia of *geomagnetism and paleomagnetism*, Eds. Gubbins, D., and Herrero-Bervera, E., Springer. This is a very valuable brief account of the magnetometer.
Collinson, D.W., 1982. *Methods in Rock Magnetism and Paleomagnetism*.
Fuller, M., 1987, *Experimental methods in Rock Magnetism and Paleomagnetism*, In Methods of Experimental Physics, Academic Press.

The latter two texts cover experimental methods in paleomagnetism and rock magnetism.

Appendix E
The Generation of Magnetic Fields in Electrically Conducting Media

The general principles of (re)generation of magnetic fields are now well established. Two processes are involved.

The first is the creation of new magnetic field from an ambient field by the motion of the fluid. This is admittedly a little mysterious. It comes about because magnetic field lines are trapped in good electrical conductors, such as the metallic fluid in the outer core. Although this trapping of magnetic field lines has important technical applications, we are not familiar with it in everyday life because the atmosphere in which we live is such a poor electrical conductor. Were it a much better electrical conductor, we would not be able to move magnets with their associated magnetic fields around so easily. The trapping follows once again from Michael Faraday's celebrated law of electromagnetic induction, which tells us that when a magnetic field changes, an electromotive force is set up in conductor, inducing a current to oppose that field change. Hence in a good conductor, such as the earth's core, any movement of magnetic field lines changing the field is strongly inhibited. The field line is therefore trapped in the medium—the frozen field effect. The magnetic field is carried along with the fluid as it responds to the forces imposed upon it. In so doing, the field lines can be sheared and new magnetic field in a new direction is created. Figure A5.1 gives an example that may help crystallize the idea by considering the motion in a shear zone of a San Andreas type fault. On the right we see the situation after the fault has moved. The result is that the field lines are sheared with the material. The effect of the mechanical displacement has given rise to magnetic field in a different direction in the shear zone. However, we should note this is the frozen flux *approximation*—the field lines can move, but this movement is small compared with the movement due to the displacement of the material.

Two cases of special interest are illustrated in Fig. A5.2. Field lines are stretched and deformed by non-uniform rotation. Consider a single field line the fluid outer core of the earth (Fig. A5.2a). With non-uniform motion in the outer core the field line will be sheared as it crosses the outer core, perpendicular to the rotation axis—a toroidal field. In this model, it is the increase in velocity as one

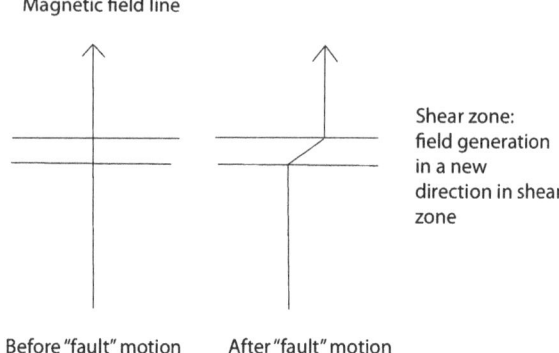

Fig. A5.1 Shearing of highly conducting material with trapped magnetic field generates magnetic field in a new direction

penetrates further into the liquid outer core that deforms dipole field lines into the toroidal fields, which circle the rotation axis. Next, rising ambient convective cells with cyclonic action regenerate field within the meridional plane field. In addition, cyclonic action on the circular field lines will again deform the field, giving a new component of magnetization. Such motion can arise from simple thermal convection, or by the rising of low density fluid, as the inner core formed. To help visualize this note that as the fluid rises the motion induced is equivalent to a corkscrew motion twisting the field lines as they are buoyed up, but in opposite senses of motion in the two hemispheres. To help further, the atmosphere is an excellent analogue. Neglecting the magnetic effects in the core, its motions are governed by the same equations as motion in the atmosphere and similar flow patterns are found.

The second process taking place in the core is the diffusion of the magnetic fields. It is at least superficially less puzzling than the first. Just as colored dye in a swimming pool will soon diffuse throughout the pool, so magnetic field lines diffuse throughout the outer core. The effect called upon here is the growth of the scale of the features seen in Fig. A5.2b to regenerate the initial dipole field lines. Yet, this diffusion must take place against the frozen field effect, which we have just discussed. It is the balance between these two competing processes, which determines the time dependence of the magnetic field, whether the field decays away, or is regenerated. On the large scale of astrophysical, or planetary bodies the field lines are caught up in the fluid motion, distorted and generate new magnetic field before they diffuse away. In the earth's core, the natural decay time of the magnetic field to diffuse away and disappear seems to be around 15,000 years. It is clear that some form of dynamo, or generator in the American usage, operates in the fluid outer core of the earth. One such model was illustrated in the figure. Recent numerical solutions of the equations for field regeneration field have been made which simulate many aspects of the behavior of the field including spontaneous field reversals. However, as is often the case in geology and geophysics, it is easier to develop models explaining nature, than to demonstrate that a particular model developed is what actually happens in nature. The details of the regenerative processes, which maintain the geomagnetic field are obscure and

Appendix E: The Generation of Magnetic Fields in Electrically Conducting Media   125

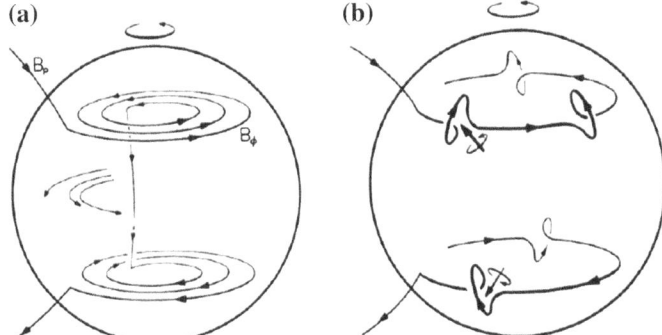

Fig. A5.2 A dynamo scheme regenerating magnetic field **a** generation of toroidal field circling the rotation axis, and **b** poloidal field which diffuses to regenerate diple field lines of B_p

unless some observation can be made that cleverly distinguishes between possible models, the details may, as Walter Elsasser once noted, elude us for ever.

Note

1. Courtesy of Peter Olsen, Johns Hopkins University many thanks.

Appendix F
Theory of TRM

Néel takes as a starting point the idea that the presence of a magnetic field adds an additional energy term to the total energy of a magnetic particle. This term is minimized, if the particle's magnetic moment is aligned with the field. To illustrate his ideas, we now consider the case of a sparse distribution of small randomly oriented needles of iron, which may not be too bad a model for some lunar samples. The shape of the needle determines the long axis as easy, or preferred, in the absence of a magnetic field, i.e., there are two equivalent preferred directions. The magnetization can be in either sense, but it must be along the axis. There are then three energy terms competing to control the magnetization of the particle.

Thermal energy (E_T) + Magnetic field energy (E_H) + shape energy (E_S)

Immediately below the Curie point when the particle becomes magnetic, it does not carry stable remanent magnetization because thermal energy is still so strong that it dominates and randomizes the magnetization. As the temperature falls and the magnetization increases, the energy due to the field first becomes important, and so the magnetization has a statistical bias in the magnetic field direction. Eventually, the energy giving the two equivalent easy directions dominates. Even when this is achieved the particle does not immediately carry stable remanence— the thermal energy is still too strong. However, in the presence of a magnetic field, there will be a statistical bias now in favor of magnetization along the axis in the direction nearest to the field. To be anthropomorphic, we may say that particles are happier in the lowest energy state with their magnetization closest to the field and therefore stay longer in that state than in any other. As temperature continues to fall, the thermal energy driving the switches of the magnetization between the two axial directions decreases and the energy minimized by having the magnetization along the long axis of the needle increases. The result is that the bias in the directions of magnetization along the axis of the needles nearest to the field direction becomes fixed as a record of the field.

As Néel's theory demonstrated mathematically, this transition is sudden, so that in a fall of temperature of a few degrees the magnetization of a particle changes from switching backwards and forwards along the axis in fractions of a second to a stable state, which can last for billions of years. We call this the blocking temperature of thermal remanent magnetization (TRM) for a particular particle. This process of TRM is more efficient than the DRM of sediments and in the geomagnetic field gives more like ~ 1 part in 100 of the maximum remanence that a sample can carry.

Epilogue

I hope that I have been able to communicate to the reader some of the joy and excitement of the adventures of the Apollo days. Many of us who worked then are now retired and no longer active in laboratories, but younger scientists are renewing the effort with more sophisticated techniques. As a result of their efforts, it now begins to look as though we have a reliable observational base fpr lunar paleomagnetism.

The early troctolite at 4.2 Ga,[1] the melt rocks and melt breccias at $\sim$3.9 Ga,[2] and the older mare basalts at $\sim$3.7–3.5 Ga[3] all appear to have acquired remanent magnetism (NRM), as they cooled in the presence of the ancient lunar field. This field at $\sim$3.9 Ga may have been as much as roughly twice the surface field on earth. It appears that the melt rocks and melt breccias recorded the field, as they cooled in ejecta blankets. The younger basalts at $\sim$3.2 Ga do not carry as convincing a record of the field and this may be evidence of the decay and final failure of the dynamo at $\sim$3.5–3.2 Ga[4] but more work needs to be done to clarify this. Similarly more data are needed to determine when the dynamo turned on. One additional aid that would make a big difference would be to have some oriented bedrock from lavas and in place ejecta blankets. This would then rigorously test the suggestion of an axial dipole form for the field.

---

[1] Garrick-Bethell, I., Weiss, B.P, Shuster, D.L, and Buz, J., 2009, *Early Lunar Magnetism.* Science, 323, 356–359.

[2] Lawrence, K., Johnson, C., Tauxe, L., and Gee, J. 2008, *Lunar paleointensity measurements: Implications for lunar magnetic evolution.* Physics of the Earth and Planetary Interiors, 168, 71–87.

[3] Shea, E.K., Weiss, B.P., Cassata, W.R., Shuster, D. L., Tikoo, S. M., 2012, *A Long-Lived Lunar Core Dynamo*, Science, **335**, 6067, 453–456.

[4] Tikoo, S.M.; Weiss, B.P.; Grove, T.L.; Fuller, M.D. 2012. *Decline of the Ancient Lunar Core Dynamo*, 43rd LPSC., ABS, 2691, This abstract presents the evidence for the decline of dynamo principally from work on sample 12022.

A very promising precession model for the lunar dynamo has been proposed.[5] It suggests that as the moon receded from the earth, this precession energy that drove the dynamo would weaken and eventually the dynamo failed, possibly around ~ 3.0 Ga., although others argue that a dynamo similar to the earth's could have operated in the first billion years of lunar history.[6] Finally we may be close to ending the controversy over the lunar paleomagnetic record and to resolving the puzzle of the mysterious magnetism of our beautiful moon, but the nature of the energy driving the dynamo for this field remains an active area of debate and research.

---

[5] Dwyer, C.A., Stevenson, D.J., and Nimmo, F., 2011, *A long-lived lunar dynamo driven by continuous mechanical stirring*, Nature, **479**, 212–214.

[6] Konrad, W., and Spohn, T., 1997, *Thermal History of the Moon: Implications for an early core dynamo and post-accretional magmatism,* Advances in Space Research, **19**, 10, 1511–1521.

The manufacturer's authorised representative in the EU is Springer Nature Customer Service Centre GmbH, Europaplatz 3, 69115 Heidelberg, Germany. If you have any concerns regarding our products, please contact ProductSafety@springernature.com

Printed and bound by CPI Group (UK) Ltd, Croydon, CR0 4YY

23/03/2026

02076448-0001